# Development and the Law

Godfrey Bruce-Radcliffe is a development lawyer with extensive experience of all parts of the process. His aim in this book is to widen the perception of those professionals concerned with construction and development and dealing with the resultant asset and whose role is, sometimes quite consciously on the part of the employer/client, confined by an almost 'need to know' basis of instructions.

**Godfrey Bruce-Radcliffe** is a partner in Thomas Eggar, specialising in the development of commercial property. He is a familiar conference speaker and has written a number of books on property development.

# Also available from Taylor & Francis

**Property Development, fourth edition**
*D. Cadman and R. Topping*
Spon Press

**Understanding JCT Standard Building Contracts, seventh edition**
*David Chappell*
Spon Press

**Adapting Buildings for Changing Uses**
*David Kincaid*
Spon Press

**Dictionary of Property and Construction Law**
*J. Rostron, Linda Wright, Laura Tatham and Robert Hardy Pickering*
Spon Press

Information and ordering details

For price, availability and ordering visit our website **www.tandf.co.uk**
Alternatively our books are available from all good bookshops.

# Development and the Law

A guide for construction and
property professionals

Godfrey Bruce-Radcliffe

Taylor & Francis
Taylor & Francis Group

LONDON AND NEW YORK

First published 2005
by Taylor & Francis
2 Park Square, Milton Park, Abingdon, Oxon OX14 4RN

Simultaneously published in the USA and Canada
by Taylor & Francis
270 Madison Ave, New York, NY 10016

*Taylor & Francis is an imprint of the Taylor & Francis Group*

© 2005 Godfrey Bruce-Radcliffe

Typeset in Sabon by
HWA Text and Data Management, Tunbridge Wells
Printed and bound in Great Britain by
MPG Books Ltd, Bodmin, Cornwall

*British Library Cataloguing in Publication Data*
A catalogue record for this book is available from the British Library

*Library of Congress Cataloging in Publication Data*
A catalog record for this book has been requested

ISBN 0–415–29021–X

# Contents

# Preface

When I was invited to write this book, I thought that rather than make assumptions about what it should contain, I should first canvass views from people I knew and had worked with, and sometimes opposite, in the property industry. I was delighted that so many people replied with their views. Most interestingly, as the replies began to arrive, it soon became clear that people's views were falling into two broad camps. The first of these, and noticeably the smaller, focused on statutory and regulatory issues as affecting them in their daily work. Construction and property professionals featured almost exclusively in this grouping.

The need for this book thus became clear almost immediately, because the property industry view which clearly prevailed was that the need was quite different. All construction and property professionals receive, to a greater or lesser extent, a measure of legal instruction as part of their professional training. The planning system, building regulations and so on come to mind immediately. Of course, it might be interesting for a professional to see what someone other than his or her tutors has to say, perhaps in the context of the interaction between the various functions of the professionals involved in the development process, but for the most part it would only be repetitive. However, what clearly emerged from the canvass was the need for construction and property professionals to be drawn more closely into the development process itself, to focus on the web of legal interaction, and to be aware of their effect upon it and, in turn, its effect upon them.

One of the most telling replies I received was from Allan Chisholm, late chairman, Lend Lease Europe Limited and, with his kind permission, part of his letter is reproduced:

> One area where I feel there is at times a lack of understanding is the breadth of legal issues involved in the development process. In particular, the way any one of these can throw the project off course is not always appreciated. Equally the long delays created by the legal procedure getting into difficulty are not always understood. This can result in insufficient attention being paid to these legal hurdles in the early stages of the development programme.

I consider that, although it is not practical for property professionals, other than lawyers, to understand the complexities of the legal process, it is important to have a general appreciation of what is required and the risks involved.

The above encapsulates the purpose of the book, namely to introduce the reader to supervening and often overriding aspects of development which his professional training will not so readily yield, and to expose some further rationale for his involvement and in particular the activities and requirements of others. That he himself has certain specific skills goes without saying, of course. Much of the material may be seen as the opposite of expectations. For example, Chapter 8, 'When it all goes wrong' suggests default and dispute in building contracts, but in substance it is not, so read on. It is thus the very antithesis of most of what is taught, and views the world from the perspective of the employer and all those associated with him.

For the construction professional, 'development' may often mean, incorrectly, the construction process itself, or at best procurement. Here is a site, the architect has prepared drawings for planning purposes, planning permission has been obtained, and the professional team are now closely involved in working up the scheme. Assisted by the quantity surveyor, the developer is seeking tenders for the construction contract. More specialised professionals are engaged because of the needs of the development, e.g. environmental, mechanical and electrical, archaeological consultants, and so on.

From nowhere at all, well, actually, from the developer's solicitors, draft professional appointments and a variety of warranties appear. The professional team have not worked with this particular developer before. Why are the appointments in this form? What is the catch? The construction professional reaches for his professional indemnity (PI) insurers, or if he can afford it his own solicitors. The developer complains of time wasting, and says his funding is being held up.

On top of this, the quantity surveyor receives from the developer's solicitors a raft of amendments to the chosen form of building contract. Why are they saying this or that? Has the developer told the solicitor nothing about the project? Is the developer close enough to the construction process to know how to instruct his lawyers? The solicitor admits that he has seen some draft employer's requirements, but otherwise that he is only aware of the location of the site (because his firm dealt with the conveyancing). His firm has also advised on some awkward conditions in the planning permission and on a draft section 106 agreement, but beyond being told the contract sum and (probably without being told why or invited to comment) the desired form, he hasn't a clue (and so it is with everyone else involved), and somehow the threads need to be drawn together. If the solicitor is deprived of the benefit of his client's thinking, or that of his professional team as the case may be, chances are that his client is working on misplaced assumptions.

However, strangely, the quantity surveyor hasn't asked how it was he was instructed in the first place (and neither has any of the rest of the professional team). No disrespect to him, the fact he was instructed at all is not the issue. That was the good news. The bad news is that he is probably not being instructed by a money bags developer at all. The name is familiar and reputable, ah yes, Bloggs Limited. But what's this? Bloggs (Neasden) Limited. Same company? Answer, no. Does an association of names imply an association of legal liability? Answer, no. It may be a wholly owned subsidiary of Bloggs but it is not Bloggs. If it has £100 capital, and £100 in the bank, that's the strength of the 'client' or, again, it might be a joint venture vehicle, say 50 per cent owned by someone the professional has never heard of, and is unlikely to hear from during the entire course of the development, and again it may still be a nominal entity.

And then, he finds that he is being asked to give construction warranties to a host of people, to a bank, an institution, tenants, the local authority for whom the developer has agreed to provide some off-site facilities, but why? Within, there are provisions for step-in or novation by more than one person. Just who is running the project, where is the money coming from? If he can't get the instructions he needs or if the source of funds dries up, to whom does he turn? Where does he fit in? As a hardened professional he might heave a sigh and answer 'I know my place', but what or where is that place?

The reality is that development often involves complex relationships, most of them contractual, to which the construction process is beholden, but of which it also forms part. More than this, the elements of the construction process have an important role in shaping those relationships. From beginning to end, as between the players, development is essentially a contractual exercise. For developers, the process is seen as top-down, from those holding the aces, i.e. landowners with an enduring interest in the built outcome, or those holding the purse strings, through those hoping to profit more immediately from the venture, to those who actually provide the construction services and the physical entity, and then lastly the occupier. If the process is viewed the other way up, the construction professional may blanch at his exposure to so many masters, but this book aims to help him confront reality, and heighten his awareness of what is actually going on, and how he may be affected.

Many thanks are due to the many people in the property industry who kindly responded to my request for thoughts as to what they would like to see in a book of this nature. Particular thanks are also due to Bill Gloyn, Chairman, Commercial Property Practice Group, AON Limited, for his help and guidance on insurance-related elements of the work.

Finally, it only remains for me to thank my secretary at my previous practice, Suzanne Merchant, for once again typing up a text for me, and to my secretary at Thomas Eggar, Sharon Moore, for finishing the job.

Godfrey Bruce-Radcliffe
December, 2004

# 1　The players

## Or: Who do you think you are?

It is only human nature, but each of the principals and professional advisers concerned with the development process regards his or her role as central. First, there are the many often faceless individuals employed by various departments of local and central government, by government agencies and so on whose role may be seen as negative, in the sense that they exist to regulate and control the development process in the interests of some public benefit which the politicians have decided should be protected. However, unless those bodies are also principals, they are not actually central to or part of the creative process.

By contrast, the principals and their professional advisers have a creative role, the contractor too. It is not enough for them simply to be identified and perform some particular task, still less perform some ritual courtship. As between themselves, they are tied into the process solely by the expedient of binding legal contract supported, according to the needs of their fellow players, by various kinds of security in the broadest sense of the term.

Throughout this book, 'development' refers to the entirety of that contractual process, from the germ of the idea through to fulfilment, and to use and enjoyment of the built result. 'Construction' is, therefore, only a part of that process. Consider it central, merely because it is necessary, and one returns to the basic misconception so often held by individual players.

In this chapter, the principal players are identified and their roles explained. How they may participate, and the likely contractual mechanisms to be used, will emerge in later chapters.

## The landowner

There is no axiom to prescribe that a landowner and a developer should not be one and the same. The landowner who builds for his own use apart, many developers will either hold or acquire land specifically for development. They may even be sufficiently resourced to engage in the construction process without reference or deference to others, bankers, joint venture partners and so on.

The presence of an independent landowner suggests two broad possibilities, first, that the developer has been engaged simply in a quasi-professional role and/or, second, that the landowner has some kind of residual or continuing in the built outcome which needs to be preserved, notwithstanding that the developer will at some point acquire a major interest in land, say a freehold or a leasehold. A lease is a convenient mechanism for the effective imposition, for example, of future management and the collection of charges for services, or the general regulation of the development over time and in a secure manner. A freehold, by contrast, is precisely that and is not subject to the ultimate sanction of forfeiture. Conveyancing solicitors will happily point to anomalies such as the estate rent charge, being the only survivor of the genre since 1977 (Rent Charges Act 1977), breach of which may give rise to a right of distress. Such niceties are perhaps best left to conveyancers.

A landowner need not be a principal in the development process if he is simply concerned with future use and enjoyment of the land. He may well be able to secure this simply by a restrictive covenant, that is to say a covenant, negative in substance, which, if the burden is passed to successors, must also be shown to benefit the landowner's other land and relies for its effectiveness on being registered at the Land Registry. Again, the legal rules relating to restrictive covenants as they affect unregistered and registered land are best left to the conveyancers. Restrictive covenants can also be used, for example, as a mechanism for clawback, that is to say, securing some financial benefit as a consideration for release of the restriction in question [Shiloh Spinners v. Harding, House of Lords 1973].

The practicality and the prospect, however, is that the landowner as a player has some interest in the built outcome which requires regulation from the outset. This implies the imposition by him of milestones and performance criteria, of which timing is but one of many. Accordingly, every other relationship, e.g. by the developer with funders, tenants and so on, professionals too, is essentially subservient to it. For example, the landowner wants works carried out on other land perhaps wholly or partially in lieu of monetary consideration, as a condition of grant of the developer's interest in the development site, and before it may be occupied for commercial use. Or, perhaps the landowner has a more tangible interest in the principal development, e.g. a share in prospective rents, or control over future use as part of its strategy as, say, a local authority. It may be that the development is in other ways critical to the use and enjoyment of its retained assets, e.g. a railway station or transport interchange, perhaps part or the whole of which is to be rebuilt as part of the commercial development. This is often more convenient than simply taking cash, particularly if the chosen developer has relevant and valuable skills.

It behoves the developer to negotiate his commercial agreements with the utmost care, in the knowledge that unless the other players in the piece are already part of that process, he will be judged on the success of that negotiation by their willingness (or unwillingness) to participate in line with the agreed terms. This reaches right down to the appointments of the professional team some of whose work may have been commissioned as a precursor to, but most of which follows, the event, and ultimately to the building contract itself. The resultant development agreement may feature large in the eventual building contract and perhaps be reflected in particular provisions of professional appointments.

Contractors and professionals, particularly those professionals who are administering the building contract, should take care when a development agreement between the landowner and a developer is in some way woven into the fabric of their own obligations. There are times when a professional may be concerned to receive legal advice as to the prospective impact on his own role. This may reflect a certain timidity, or it may point up the need for the developer to be absolutely clear in his requirements. It may also point up the need for the lead professional, say a project manager, to be up to speed on the requirements, and if the written contractual relationship between the developer and his other professionals prescribes a measure of delegation of authority to the project manager, the professional must in turn be clear in his mind on what and whom he can rely, and whether and how far he must follow his own judgement in interpreting those requirements. Happily, it will often be that what is required is plain on the face of it. If it is, or seems, obscure, then clarification should be sought: misunderstanding or allegations of misrepresentation have a potential all of their own.

Finally, it is rare in commercial development to find that individual players *per se* are flesh and blood human beings. They may be public bodies with statutory corporate identities such as local authorities, regional development agencies and so on, or they may be corporate entities. Sometimes they will be companies providing the interface for an underlying joint venture, possibly a partnership between corporate entities, a limited partnership or a limited liability partnership, all of which have unique legal characteristics. In the day-to-day conduct of development, the (very real) distinctions may not loom large, or at least appear to do so, and therein lies the danger. If there is default, or a dispute, then the constitution of the particular player and the security it has given to others, of whatever kind, may be critical in determining the ability of that player to perform.

Later, we shall consider the effects of constitution of the developer, indeed any contracting party, on legal liability, including the doctrine of *ultra vires* (beyond the powers). Suffice it to say at this point, however, that no assumption should be made of a player's ability to perform. The most common constraint is a natural one, resources. The simplest notion is that of a company which has cash in the bank unclaimed and unsecured. Now consider a public-sector body such as a local authority or an agency of government. There is, actually, no bottomless pocket to speak of save ultimately the embarrassment of central government. In the 1990s, real problems were encountered with the advent of NHS hospital trusts as principal players in hospital-building schemes under the private finance initiative (PFI). First, there was the issue of *ultra vires* which required special legislation so that a project could be certified by the Secretary of State [National Health Service (Private Finance) Act 1997]. Second, there was the problem of residual liabilities if, for example, an NHS hospital trust were wound up and its assets otherwise taken back into whatever body was deemed appropriate at the time, e.g. the local health authority [National Health Service (Residual Liabilities) Act 1996]. Residual liabilities are thus now covered, but when dealing with all agencies and public-sector bodies, a lawyer needs to go directly to the enabling legislation to understand his client's position. The fact is that, unless something is expressly prescribed, e.g. guaranteeing by government of borrowing (and the guarantee is actually given), there is no implied mechanism for deepening the pocket. One thus has to rely on the willingness of politicians to commit funds before the bottom is reached. Fortunately, the UK is not bankrupt yet.

## The developer

A resourced developer may most often be expected to be a single corporate entity. It may be the creature of financial interests other than direct funding, e.g. venture capitalists or joint venture partners of whatever origin, and thus the subject of some underlying form of joint venture.

It is therefore essential to understand something of the nature of the corporate entity. It is a legal person, and exists as such because it has been created under the Companies Acts and duly registered as a company at the Companies Registry. As a non-human legal person given life only by statute, it suffers from potential constraints of *ultra vires*. Public-sector bodies as well often have some measure of relief in their statutory constitutions, particularly in the case of disposal of assets. It is not a total comfort, however, and the same applies to companies.

The objects for which a company exists are contained in its memorandum of association, and the regulation of its affairs is governed by its articles of association. Activities beyond prescribed objects are *ultra vires*. There may also be a shareholders' agreement further regulating its affairs as between its own players. If this binds all its shareholders, it too may be registered. Whilst the Companies Acts provide a measure of protection in terms of actions approved by the board, *ultra vires* is not an entirely dead subject, and if lawyers should not make assumptions, neither should others.

That said, however, issues of *ultra vires* concerning companies are infrequent: the more likely concern is one of financial resources. Whatever the constitution of the developer, resources are fundamental and a specially created subsidiary, perhaps with part of the name of the proposed development or of the immediate holding company in its title, may be an entirely nominal entity. Witness the recession of the 1990s which left so many professional fees unpaid and professional firms bankrupted. We have had recessions before, such is the pattern of commercial life, but the 1990s recession was also endured post-Insolvency Act 1986 which created new kinds of insolvency control to protect insolvent businesses and to preserve the prospects of revitalisation, however remote in practice. It is as well to mention certain of them here because it is the influence of that act which today characterises the nature of the 'developer'.

The first of these innovations was 'administration' under which creditors could apply to the court for an administration order. If successful, this would result in a moratorium on enforcement of indebtedness whereby, if a means could be found to restore the company to solvency, the appointed administrator would work to that end. There have been some notable corporate administrations, not development-related as such, but in terms of numbers they are dwarfed by the other major innovation, administrative receivership, which has affected many a development. Without doubt, the influence of the Insolvency Act 1986 has gone to the root of development and in particular the identity of the developer.

The essence of administrative receivership is (or rather was – see the Enterprise Act 2002 below) that so long as the bank or lender has security over the whole or substantially the whole of a company's assets and the borrower (mortgagor) is in some kind of breach whereby the security can be realised, then an administrative receiver can be appointed to run the affairs of the company (and he may, in particular, or rather could oppose and thwart the appointment of an administrator by the court, and thereby the imposition of a moratorium). With these new-found powers, the banks promptly maximised their security. No longer was it appropriate for a developer to run a series of developments through one principal company, each secured at different banks to the extent of the development in question. Now, in practice, a separate subsidiary would need to be created, a single-purpose vehicle (SPV), for each particular development, with a debenture to be given over all of its assets (such as they might be), thus placing the bank in control. And so were born the nominal corporate entities masking as 'developer' (e.g. Bloggs (Neasden) Limited – see Preface), whilst the bank, while lending to it, could look to the covenant of a separate guarantee from its holding company (e.g. Bloggs Limited). However, important changes were made by the Insolvency Act 2000 and the Enterprise Act 2002, which are now in force.

It is helpful to outline these changes in the law because they permeate the default provisions of every commercial contract including leases. They therefore also affect a professional's relationship with his corporate employer in times of the latter's financial distress. The Insolvency Act 2000 provisions, here discussed, apply only to small companies as that term is understood by Section 247(3) Companies Act 1985. That is to say, the company has a turnover of not more than £2.8 million per annum, a balance sheet total not exceeding £1.4 million and not more than 50 employees. To qualify as a small company, the company must satisfy two or more of those requirements. Essentially, in case of insolvency even post-1986, a company voluntary agreement with creditors would inevitably invite hostility from debenture holders, and floating charges would crystallise. Sections 1 and 2 of the 2000 Act and Schedules 1 and 2, now provide a choice for directors of a small company who may proceed with a voluntary arrangement, with or without a moratorium which, unlike post-1986 administration, does not require an order of the court. The use of such an arrangement, therefore, obviates the need for formal administration which, as before, still requires an application to the court.

Procedure apart, a statement of the company's nominee (who does not have to be a licensed insolvency practitioner) filed in the court, in so many words that the arrangement has a reasonable prospect of success, facilitates the imposition of a moratorium preventing enforcement by a secured creditor of its security without the permission of the court. Particularly, debenture holders are precluded from applying to the court for permission to enforce security as they would have done under the previous law. Previously, a debenture holder could effectively override an administration by appointing his administrative receiver, and crystallising the floating charge, but no longer can he do so pending termination of the moratorium. Members and creditors can apply to the court in certain instances of default, during the moratorium, the court having wide ranging powers.

The moratorium itself can last up to 28 days although it can be extended by a further two months by resolution after meetings of the company with its creditors. It will, therefore, be seen that this is only a temporary measure and that, eventually, the debenture holder will prevail. However, it should be borne in mind that this out-of-court moratorium procedure, available in addition to formal administration, is intended to apply only where there is a reasonable prospect of the moratorium enabling the company to carry on its business. If that is not the case, debenture holders will have their way.

However, consequent upon the Enterprise Act 2002, the position of debenture holders has changed, as has also the whole fabric of corporate insolvency administration. The relevant provisions do not affect existing floating charges under which administrative receivers can be appointed as before. Remember that where the security constituted the whole or substantially the whole of a company's assets, the floating charge holder (debenture holder) could appoint an administrative receiver so long as the security covered the whole or substantially the whole of the company's assets. A floating charge is distinguished from a fixed charge in that the assets secured can be dealt with by the company until the charge crystallises through an event of default as prescribed by the charge. A person dealing with the company needs to be satisfied that such security has not crystallised, best done by the charge holder itself certifying to that effect. A director of the company can certify and for most purposes that is usually acceptable.

The main thrust of the new legislation, as regards administration both of companies large and small, is to obviate the need for a court order, but relevant papers still have to be filed in the court. In most cases, administrative receivership (including of larger companies – but see the Enterprise Act 2002 below) will not be available to a debenture/charge holder, giving way to a single administration procedure. Given that the chargeholder once had to intervene in a court application to appoint an administrator, in order to appoint an administrative receiver, the appointment of an administrative receiver did not require a court application as such. The new procedures will, thereforc, suit charge holders perfectly well, save that they will be operating under a new regime. As it is early days, it remains to be seen if any flaws will, in the event, manifest themselves. As to the actual procedure, it is helpful to understand a little of how this will operate. A bank lender intervening will soon make his presence felt amongst all those concerned with the development. The new law applies to most companies save where the indebtedness exceeds £50 million and the issue of a capital market investment. Administrative receivership will also still be available to a project company where the project is a public–private partnership involving step-in rights. A further exception is where the company is a registered social landlord under Part 1 of the Housing Act 1996. The Secretary of State is further empowered to add or remove categories of exclusion.

These exclusions apart, administration, using the new non-statutory procedure, will thus be available where rescuing the company as a going concern is not reasonably practicable. There are important drafting points for lawyers to ensure that floating charges qualify for administration, however. An immediate moratorium arises once notice is given by the company or its directors of intention to appoint an administrator and notice must also be given to a qualifying floating charge holder which has five business days in which to appoint its own administrator if it does not consent to the choice. If the floating charge holder does not appoint its own administrator the company or its directors can proceed with their own administration. Of course, the floating charge holder itself can initiate the procedure.

As has always been the case, those finding that they are dealing with an insolvency situation should take legal advice immediately. It should be remembered that a bank lender is concerned, having once lent, with being repaid but, unless there is an economic downturn, one of the best ways of achieving this may be through allowing reconstitution of the scheme in the hands of a new developer who may well want to appoint the original team, assuming that they were not the cause of the problem in the first place. A landowner under a development agreement (see Chapter 3) will see developer insolvency (and so prescribe) as an event of default, hence the need as will be seen for an accommodation of the bank in default procedures in this respect.

Moving on from insolvency, there is no rule of law whereby a holding company is in some magical or mysterious way inherently responsible for the debts of its subsidiaries. Prospective liability of directors for trading while insolvent apart (necessitating an application to the court by the liquidator) unless one has the guarantee of the parent company or at least of a company with sufficient assets, the covenant of a company is essentially as good as its own assets and no more. Time and again contractors enter into building contracts through (SPV) developer subsidiaries, with no thought for the contractual consequences. Professionals are often no different in their approach. Finance directors can be expected to know better, but unless resources are contractually accessible, the developer may be no more than a house of cards (and tends to be precisely that, because of Insolvency Act and subsequent requirements).

The nature of the entity apart, another concern for the players at large is the skill of the developer. Sometimes developers are peopled by highly experienced professionals who are hands on, and are prepared to have a robust working relationship with all the other players, including, not least, their professional team. That is an ideal which is far from always present. Where the technical skills are lacking but the imagination intact, the logical fallback is to appoint a professional person or firm to fulfil that role. This is particularly important for businesses who make no pretence of having development skills and who rely on professional help in relation to projects with which they are directly involved. Indeed, such is the role of the project manager in such circumstances that this may influence the entire structure of the construction process.

Perhaps the most dangerous kind of developer is the wheeler-dealer who may have a trusted professional team, but whose personal lack of technical knowledge or business skills means that the team have thrust upon them an additional burden of consultation with other members and of seeing through the project to fruition. Many architects are excellent contract administrators, but it is unfair to visit upon them (and perhaps even more unfair to fail to visit upon them) the underlying commercial exercise of which the developer has perhaps only a limited grasp. This author, having practised as a development lawyer for over thirty years, has experienced all these kinds of developer. A simple lesson emerged early on for this author: as a lawyer, whoever your client, both draft and amend development documentation with the client's protection in mind. Never compromise that standard except on the strictest instruction. Strive if you can to cater for the barmy or unconscionable, but never be surprised if you are caught out. Advice to other professionals is therefore, at all times, be circumspect.

## Joint venture partners

At this stage, read nothing into 'partners', partnership being a particular legal structure having particular legal characteristics, some of which will be discussed later. A multiplicity of players suggests a more complex structure than the age-old triumvirate of landowner, developer and funder. If the joint venture is represented by a single corporate entity then all the interests of the players within that joint venture are brought together in one legal person, a single-purpose vehicle, therefore, but it may still be just a nominal entity, possibly with each of the players itself using an SPV to contain exposure. Those involved in the construction process are, nevertheless, beholden solely to that single entity (but may have the benefit of guarantees from one or more of the participators or their wealthier parents). The prospect of separate responsibility to one or more joint venture partners is perhaps remote unless that responsibility is necessarily unique to the player or players in question.

There are many possible reasons for the existence of a joint venture, perhaps the pooling of skills and resources, risk capital and so on are all potential candidates. Later, we will consider some forms of joint venture which are commonly to be found in the development process. The most recently contrived entity, the limited liability partnership, has, because of its novelty, and also certain tax constraints yet to be seen to proliferate. Whatever the legal structure of a joint venture, joint venture partners will seek to contain their exposure (particularly if there is a true partnership in legal terms, about which more anon) by themselves creating SPVs. Accordingly, whatever the legal structure of the joint venture, and however well resourced the ultimate players, it may still be a nominal entity, and probably will be. Those who rely on the comfort of financial resources must, therefore, consider carefully with whom they are actually dealing and seek the comfort of appropriate security, most usually guarantees. It is the funders, however, who will be seeking primary security, on the land in particular, and all other interests tuck in around and behind them.

## Funding partners

Development finance is a complex subject. If not also participating as a joint venture partner, funders may be seen as falling into two broad groups, lenders and investors. Risk capital (in one sense, all capital is risk capital) is another matter.

The wheel has been invented many times over, but the provision of development finance has certain well-worn characteristics. All financial support has a measure of speculation in it: nothing is totally secure. (The nearest thing to security in this life is a government stock.) Lenders are in it to provide some return for their depositors. Investors are in it to provide some return for themselves or their beneficiaries. All are in it for themselves in some form. Thus, an investor is looking to acquire an asset with the long term in mind. A hybrid concept, perhaps less often seen in practice (although common many years ago), is the investment mortgage, providing a long-term loan. More likely, however, a lender is there for the short to medium term, to ensure that his money is spent wisely, that there is a take up of the built outcome, the creation of sale value and, very particularly, the repayment of his loan.

Both lenders and investors are consumed by timing (tenants too). In the case of a lender, interest must not run for so long that the economics of the project so erode the return to the developer that the loan itself cannot be repaid. Conversely, unless there are timely draw-downs, interest does not run and funds are not being put to work. The availability of draw-down must assume take-up within a period, and loan agreements may prescribe trigger payments in case of late take-up. The loan period must be finite to facilitate withdrawal of deposits and recycling into fresh projects. Mainstream loan finance is thus characterised by loan agreements or facility letters, backed by a raft of security reaching down to all of the contractual relationships which the developer holds with those actually carrying out the development, not least, those concerned with the construction process.

It should also be noted that, in the event of a lender having to take control (there will always be a measure of supervision, not least through the medium of performance criteria), banks and lenders exist for the purpose of lending, i.e. lending out of customers' deposits, and not for the purpose themselves of performing the role of the developer. Sometimes through the medium of receivership (let us assume, administrative), the business of the SPV developer may be wound through to its final conclusion. The earlier in the process, however, the more likely the lender will look in the alternative to finding a fresh developer (who may be funded by someone entirely new) as a means of realising its security and making a dignified exit. As will be seen later, land-owners entering into development agreements with developers must cater for the need for a lender to be able to realise its security and the possibility, however distasteful and unconscionable, of a new developer. There are, accordingly, certain elements which must be incorporated into any development agreement to make it 'bankable' (see Chapter 3, 'Development and forward funding agreements').

Investors, by contrast, have an entirely different perspective. In their relations with a developer there will inevitably (indeed there must) be a certain common thread. That common thread will be found in landowner/developer agreements, in loan agreements and investment funding agreements of their various kinds. A banker, on the one hand, will wish to see a development let and occupied in order to maximise capital value and ensure that its loan is repaid. An investor is similarly concerned but, instead, in order to enhance the investment value. Both are concerned with timing, in the case of the investor to ensure not only that he is creating value and income, but also that his funds are committed within a certain time-scale. Failure to achieve this means a shortfall in return to his beneficiaries, perhaps a pension fund, the general investments of an insurer, or simply an investment company. Whereas the investor is essentially interested in the completed and let development, the lender is interested in completing and letting it, and having his loan repaid.

At some point, those involved in the construction process, including professionals, will find themselves involved in commitments to a number of parties with a financial interest of some kind in the project. There may be more than one lender, there may even be a consortium of lenders (often through a lead bank), or there may be a consortium of investors. Small wonder, therefore, that construction professionals can become alarmed when they are presented with a raft of warranties. What has to be understood is that there is a sequence of commitments with characteristics of their own which make the project bankable.

For example, construction warranties contain provisions for assignment. There are often concerns about the number of assignments and to whom the benefit can be assigned. It is helpful to reflect the fact that the value of a warranty in any person's hands is limited in law by his interest in the development. If he cannot lose through lack of relevant interest, what, therefore, can he enforce? Damages follow loss or damage: in our common law the notion of imposition of a penalty in a contract is void. Whilst there may be inhibitions on assignment none the less (see Chapter 7, 'Construction procurement in the development process') assignment to a lender must usually be considered as freely available. If a construction professional or a contractor seeks to qualify an assignment provision in favour of a lender (mortgagee), all he is doing is making the project unbankable, and it is only a foolish developer who will allow it. In any case, the bank will properly disallow it. Other constraints may be disallowed as well (again, see Chapter 7), and on personal experience one has from time-to-time been faced with a funder requiring that unless the terms of appointment (or as the case may be a warranty) of a particular construction professional meet certain requirements, the developer shall be prohibited from employing him at all. (And on one experience in particular, encouragement of the funder to take a particular line proved a useful negotiating tool for the developer; devious may be, but so effective!) Long-term relationships, perhaps over a number of projects, can be severely strained by others' requirements.

Where funding partners are involved, the developer is thus constrained. It matters not how close the relationship between the developer and members of his team, the developer is necessarily beholden to others, and it is part and parcel of the contractual interaction which flows through the development process from top to bottom.

## Tenants

For the landowner, developer (including joint venture partners) and funders – unless the project is for own occupation – tenants are a vital ingredient. They are the *raison d'être*. They provide income and thereby create value. In our economy, land, bricks and mortar are otherwise a liability, not in the sense of being on the wrong side of the balance sheet, they are after all 'assets' so far as they go, but because it is only demand for them that adds value. It goes without saying, therefore, that principal development agreements, including funding agreements of their several kinds, include provisions for letting, both as to identity of tenants, suitability and commercial terms (unless everything is predicated or conditional on a pre-let). It is to be expected, therefore, that there will already be pre-lets on terms acceptable to principal players and, insofar as there are not, the development is said to be speculative. In the golden age of the 1980s, such was the confidence of banks and institutions, landowners too, that speculative development was an everyday event. Prospective tenants' requirements had to be anticipated of course (they always must be) and, insofar as they were not, one had to accept that relationships might need to fall in line in order to secure the tenant. So it was, and so it always will be.

However, just for a little history, the assumption of lettability made by banks and institutions was soon shattered by the recession of the 1990s. The shift from a landlords' market to a tenants' market, insofar as there were tenants available at all, meant that letting criteria, minimum rents and so on, as prescribed by development (funding) agreements, could not be met. The construction industry and so many professionals involved in development could only look on and let events take their course. Just how secure their contractual relationships were was soon tested, and many professional firms in particular failed, not least architects.

With a more realistic attitude to letting since, the prospect for the most part today is that pre-lets tend to be the norm. A well-advised tenant will be looking to tap into the whole of the development process, construction included, for reassurance as to the delivery of his premises. He is concerned not just with quality (performance criteria) but also with timing. He needs to open and trade, and if the development or his part of it plus related facilities and infrastructure cannot be delivered within the anticipated time frame, there is a point when he must withdraw. He will seek a commitment from the developer in this regard and, under the relevant professional appointments, the developer will likely seek a commitment from construction professionals which thus by implication flows through to warranties. Every player is affected by timing and it is in the interests of all concerned to limit their exposure. As will also be seen later, consequential loss in contract (economic loss in tort) are to the foremost in the minds of those providing the building and providing the service, and ways and means must be sought to curtail or ring fence that exposure, including through the medium of insurance.

## Public and quasi-public sector players

It will often be that a public-sector body will emerge as a player, particularly in the landowner capacity. Development commissioned by local authorities, e.g. town centre developments, and by railway undertakings (Transport for London (TfL) and Network Rail – formerly Railtrack – in particular) all have a statutory base from which to perform their functions. The advent of PFI in the 1990s gave rise to the provision of privately funded development for the benefit of the public sector, e.g. NHS hospital trusts. In a book of this nature there is little need to expand on PFI. Granted that there are time-consuming rules of procurement, by the time the exercise reaches down to those providing construction services of their various kinds, the participators are competing for the business as much as they would with the private sector (and if it is PFI they are, in the sense that it is the private sector that provides).

There is also a certain mythology about the nuts and bolts of PFI, which attempts to distinguish itself from conventional procurement by emphasising the service element in the provision of the facility, say a hospital. It may seem alarming to the non-lawyer but, to a lawyer, a contract is a contract is a contract. A label is a label is a label, and no more than that. The sole issue is what you are trying to achieve, how to express it contractually and how far it should be secured, and if you are a lawyer, the applicable law and how to express the relationship. Rules and notions of procurement apart, if notwithstanding what you end up with is still a development agreement of sorts, then so be it.

Some public-sector bodies, however, have been set up precisely with the intentions of promoting development where needed, under the general banner of 'regeneration'. This has manifested itself in a number of ways. For example, enterprise zones were initiated in the early 1980s with the view to attracting development into defined areas (Local Government Planning and Land Act 1980) through capital allowances, rating relief and so on. They had a limited life (ten years) and, strange to say, there are still one or two around to this day. Urban development corporations were also set up with the view to regenerating specific areas.

In the early 1990s, no doubt with a crippling recession adding fuel to the fire, the government of the day created a body called The Urban Regeneration Agency whose statutory basis was tacked onto an otherwise unrelated act, The Leasehold Reform, Housing and Urban Development Act 1993. Section 158(1) of the act intoned 'There shall be a body corporate to be known as the Urban Regeneration Agency …', and so they called it English Partnerships! This was designed to be an engine of regeneration having, at least in theory, a multiplicity of powers to promote regeneration, to provide finance, and even, in certain circumstances, to exercise compulsory purchase powers, and also have certain planning and highway powers, if necessary, to override local authorities. All of this was subject to intervention by the Secretary of State. To those who worked with it, English Partnerships (EP as it is sometimes shortly known) could at times seem highly bureaucratic and lacking in enterprise, and perhaps no more effective than its statutory predecessor English Estates.

The latter half of the 1990s saw another approach, through the creation of regional development agencies (RDAs), the functions of English Partnerships gradually being subsumed into these. The enabling legislation (Regional Development Agencies Act 1998) shifted the emphasis, which can be summarised very simply, in that land regeneration would give way to economic regeneration, and powers would be exercised, not so much on the basis of failure of ministerial intervention but consequent upon ministerial intervention. The politics are far too subtle, but those who deal with RDAs on property-related matters find that, in the main, they are dealing with the same people, if not the same process. Certainly, from a development perspective, it may be argued that there is little to choose, and whether regional accountability has been achieved thereby can be argued on another day.

However, as English Partnerships' natural statutory successor, RDAs are a conduit, amongst other things, for the dispensing of grants emanating from regional government offices. These combine related ministries, such as the DTI, and are also a focus for application for a variety of grants, e.g. single regeneration budget, and also the dispensation of European grant funding. Once grant funding, of whatever kind, is awarded, RDAs have an important role in administration. It is beyond the scope of this book to discuss the grant system. Suffice it to say, here, that developers and businesses requiring premises or wishing to relocate for other reasons, should approach regional government offices to ascertain the location and extent of assisted areas and the grants available to them. Far be it from this author to recommend to development professionals the art of 'ambulance chasing', practised by his distant colleagues in other disciplines within the legal profession, but all those involved in development should note and enquire.

Finally, when dealing with public-sector bodies, one is dealing with statutory corporations. They exist and function solely by and through their statutory basis. Very occasionally, the issue of *ultra vires* arises, particularly in relation to financial commitments and examples of this have been highlighted earlier in this chapter. Competition for public-sector work is hedged around by complicated procurement procedures, albeit tempered by 'best value'. One should expect the resultant terms of engagement, if direct by the public-sector body concerned, to be complicated, albeit not always, and when in doubt one should take advice.

# 2  Joint ventures

## Or: Can I join in too please?

Having identified the players, one is tempted to rush on immediately to the various kinds of development agreements which, in turn, prescribe (amongst many other provisions) how the professional team are to be appointed, and the criteria and general tests which must be met. However, between all this is the matter of joint ventures. Indeed, one or more of the principals may itself be the product of a joint venture. The purpose of this chapter, therefore, is to identify some of the more common forms. It may also help, in case of conflict (not of interest, but by way of default on the part of, or dispute with, a principal), in assisting one's understanding of 'the nature of the beast'. In other words who, exactly, is one dealing with and, perhaps crucially, who else may be accessible from whom to seek redress?

It is tempting to make assumptions about resources and accountability, but reputation may ultimately count for nought if a particular player, however it is constituted, has so arranged its affairs to minimise its exposure. That, unfortunately, is the name of the game as it is the lawyer's task to contain his client's exposure to the maximum extent. At all stages, and questions of taxation apart, it is minimisation of risk that dictates the structure.

### Co-operation agreements

Insofar as relevant to development, at least one of the parties to a co-operation agreement may be a principal player. The contractor and professionals may be wholly unaware of the existence of such an agreement unless perhaps two or more of the players are party to it or it impinges directly on them.

Before we get too deeply into different kinds of agreement, particularly in Chapters 3 and 4, it is helpful to understand a little bit about the nature of contracts which, in general terms, do not actually have to be in writing, if in the form known technically as 'simple' contract. There may, actually, be nothing simple about it at all, and a contract in writing signed by the parties to it (or executed, which has a particular legal significance of its own, as will be seen) thus evidences the detail.

Simple contract and 'covenant' both rely upon ancient common law doctrine for their legal effect. Simple contract relies on 'consideration', in that an obligation by one party is matched by the obligation of the other, thus creating a binding contract. Sale of goods is but one example. Consideration need not be adequate, indeed it may be heavily imbalanced, but it must be sufficient in law. A substantial obligation may be secured by a consideration of as little as £1, for example, as is often to be found in commercial agreements.

A simple contract in writing is characterised by being 'under hand'. However the contract is expressed, default is not actionable indefinitely. For this, one looks to statute which prescribes that in the case of a contract under hand, the limitation period runs for a period of six years from breach of the obligation. (The rules of limitation for tortious liability operate differently and the detail is beyond the scope of this book. However, in general terms the period is six years from when the 'cause of action' accrued, modified significantly by the Latent Damage Act 1986 which is particularly important to development in that it shortens the period in certain circumstances and imposes an overriding limit.)

By contrast, a contract which is 'executed' is subject to the doctrine of covenant. Once upon a time execution was quite literally under seal but, since the Companies Act 1989, this is no longer obligatory, even on the part of individual players. The essential element is that the document be expressed to be executed as a deed. So long as a deed is executed, there is no legal magic in use of the term 'covenant' although it is still frequently used in deeds. Moreover, interests in land and other legal interests (called 'choses' or 'things' in action) cannot be transferred or granted in law except by deed. There are also such concepts as equitable assignments/grants, but the principles of equity are beyond the scope of this book. A contract for the sale of land creates an equity, for example.

The essential characteristic of a covenant is that it can stand alone and is irrevocable unless released by the covenantee. A covenant, therefore, can be entirely one sided: there is no need to offer consideration in return for a covenant and the limitation period is twelve years, and not six. Because of the nature of limitation periods, many contracts expressly curtail limitation even further. For example, lest any obligation be seen to be outstanding (as opposed to breach), construction warranties generally provide that liability shall cease on the expiry of so many years, usually twelve, after the date of practical completion. Let us now move on.

A co-operation agreement is a contract of whatever kind, and no more. It is simply a contract defining rights and responsibilities, governed by the common law of contract (however modified by statute), but not inherently based on or dependent upon statute. Each or all of the parties to a co-operation agreement are thus principals in relation to that agreement. Each has unlimited liability for his actions, and owns, uses and applies his assets, skills and so on in the performance of his obligations.

Co-operation agreements, if that is what they are on analysis (see below), are inherently tax transparent. A service element, again if that is what it is, implies the generation of a prospective charge to VAT, however. A freely negotiated co-operation agreement necessarily implies that neither party can bind the other to some obligation to a person who is not a party, but so soon as that arises, so also may a partnership which, as the term is strictly construed in law, renders the relationship subject to a body of law the avoidance of which, in substance, can only be achieved through the terms of the contract itself.

Co-operation agreements have their uses whenever the need arises. Even some forms of development agreement may, upon analysis, amount to no more than a co-operation agreement. For example, a developer buys land with the view to development. The seller of the land may defer part of the consideration to the successful outcome of the project and the derivation of profit by the developer. It is always helpful to remember that labelling a contract of any kind does not itself establish its legal status. It is the substance that counts in law, especially when issues of taxation arise. Thus, corporate and personal taxation apart, a profit-sharing arrangement may be perceived as providing further consideration (if comprised in an independent agreement, it may thus escape being so considered). Insofar as that consideration is ascertainable at the time of purchase, it will in any case be liable to a charge to SDLT (stamp duty land tax). Where it is unascertainable, and whereas the old law provided an escape route, the SDLT regime imposes what is, in effect, a wait-and-see provision (Finance Act 2003 and subsequent statutory instruments). Another taxable area is VAT, particularly if there is a service element or if the further payment or payments are deferred consideration, and the sale was liable to VAT in the first instance, then a further charge to VAT will arise.

The above examples simply highlight the fact that it is the substance, and substance alone, of an agreement which characterises the relationship between the players, their prospective liabilities to tax and, possibly (particularly if a partnership be shown) individual players' responsibilities to third parties. (Having said that, actually calling an agreement a partnership agreement may raise a presumption unless the substance clearly overrides.) Mention should of course be made of the Contracts (Rights of Third Parties) Act 1999 which permits third parties identified within its terms themselves to enforce the contract, subject to the limitations on liability imposed by the contract itself.

However, it will thus be seen that while a developer's co-operation agreement may be of no interest or concern, such an agreement might confer a benefit upon one or more of the parties to it which depends in some way upon the performance of professional obligations to the developer. The existence of a co-operation agreement may thus appear if, for example, the professional is asked to provide a warranty.

## Partnership

It is timely, therefore, to mention partnership in the formal sense. Partnerships are often used as vehicles for interested parties to development. The partnership itself may not necessarily, as a partnership, become a player in its own right. Often it will use a company in which it owns the shares, and which in consequence is part of the partnership assets, to comprise the 'developer' or other principal player in question.

Formal partnerships have long been a useful medium for persons to join together in business. They are governed by statute, in particular the Partnership Act 1890. Under Section 1 of that act a legal partnership is defined as 'the relationship which subsists between persons carrying on a business in common with a view to profit'. For a partnership to exist, no formal declaration is required. Indeed there may be no partnership agreement at all, with the unfortunate consequence that in the absence of agreement between the partners, any dispute (particularly one which may give rise to dissolution), may result in a solution which is imposed by the law as it stands.

All business relationships and contracts generally have the potential for comprising a partnership. Always it is the substance that counts and, if a partnership be found, certain legal consequences follow. First, there is essentially no constraint on any legal person being a partner, so that the partnership may comprise individuals, corporations or other legal entities such as those discussed below or anywhere else in this book. Possibly, issues of *ultra vires* may place constraints on an individual partner's ability to perform his obligations, or the extent to which they may be enforced against him.

These considerations apart, partners are deemed in law to be the agents of each other for the purposes of the business or venture for the purpose of which the partnership exists. In other words, partners can bind other partners. Of course, if as between themselves partners limit that authority then, but to the extent only that a third party is aware of such limitation (and not otherwise), that limitation applies. However, the prospects of a third party being 'on notice' are limited in practice to being expressly on notice. There is, for example, no central register of partnerships whereby their constitutions become a matter of public record, unlike limited companies. Partnerships are essentially private arrangements.

It thus follows that each partner is liable, without limitation, for the debts and liabilities of the partnership, even if such liabilities were undertaken by other partners without their authority or knowledge. Section 9 of the Partnership Act 1890 prescribes that partners are jointly liable for all the debts and obligations of the partners incurred during the period that they were partners. Section 12 of the act, by contrast, prescribes that partners are jointly and severally liable for any loss or damage arising from any wrongful acts or omissions of partners. By 'several' liability is meant that each partner is individually responsible for the whole of the liability in question. The lesson for partners in any business (including professional firms), therefore, must always be 'know your partner'.

Partnerships at large are not subject to the same publicity requirement of companies, although changes in partners should be notified. However, in the event that all the partners are corporate partners then, under the Partnerships and Unlimited Companies (Accounts) Regulations 1993 accounts must be filed at the Companies Registry.

The liabilities of a partnership and of individual partners within the partnership apart, there are also statutory rules which cover dissolution. They can be overridden by agreement, hence the desirability of a partnership deed or agreement expressly governing those matters. Otherwise, the advent of an irreconcilable dispute means dissolution terms imposed.

Partnership is essentially tax transparent, that is to say, that partners are taxed according to their individual partnership profits. Of course, their shared profits are assumed equal in law unless expressed to the contrary.

A partnership may be, in context, a desirable, indeed intended, conduit for carrying on business. It may have *de facto* existence, or it may exist by express agreement, possibly a combination of elements which depend on evidence to define the nature of the partnership. If partners are in dispute, and also dispute the very nature and substance of their relationship, how bountiful and joyous it is for the lawyers, and long may they so continue. Formal partnership agreements are essential to a properly run business.

If a partnership is not intended, it is not enough for one or more partners to declare the fact. The fact depends upon the substance. Accordingly, to avoid the underlying characteristic of joint liability to third parties, the avoidance of a partnership will be achieved by express agreement that partners cannot bind each other. Co-operation agreements and any other contracts having the remotest prospect of being construed as a partnership should thus always contain such a provision. To go on to say that the parties will each be responsible for their own tax and so on, even the denial of the partnership itself, are all helpful pointers, but the core exclusion counts above all other factors which go to exclude the existence of a partnership.

## Limited partnerships

As an investment medium, limited partnerships have their place, and those dealing with an entity which ends with the word 'Limited' should not automatically assume that they are dealing with a limited company. Limited partnerships are particularly useful as media for property investment, assisting the pooling of investment funds, and so have their place in the development process, e.g. through the medium of forward funding (see Chapter 3, 'Development and forward funding agreements') and through the development process to the creation of the ultimate investment.

Like partnerships at large, they are governed by statute, the Limited Partnerships Act 1907. Strangely, their potential has taken nearly a hundred years to develop, and they are now a popular medium for collective activity. Limited partnerships, unlike limited liability partnerships and still less like limited companies, remain partnerships as such. They have no separate legal personality.

Unlike ordinary partnerships, however, they must be registered at the companies registry. Although limited partnerships are partnerships in substance, with one important and crucial exception, limited partners cannot bind the partnership. The essence of limited partnership is that there must be not less that one 'general' partner who can bind the partnership and it is through his actions that the business of the partnership is conducted. The remaining limited partners, accordingly, have no active role in the conduct of the business of the partnership, although there is a consultation process. That is not to say they do not have liabilities.

The number of partners in a limited partnership was until December 2002 limited to twenty. This was the position which stood until 1967 in relation to partners at large. Turn up the headed note paper of any professional firm you care to think of, and until 1967 (as it happens, by virtue of the Companies Act 1967, but no matter) partnerships, both general and limited, were limited to twenty, but professional partnerships were then released from this constraint.

Under a limited partnership, the liability of a limited member is limited by the amount of capital he has contributed. The general partner's liability is unlimited. In order to contain his liability a limited partner may thus be encouraged to subscribe only very limited capital, but perhaps provide his substantive financial support by way of loan. Then if the partnership has the wit (and most do) it will ensure that the partnership agreement also provides that loans will be written off in case of insolvency of the limited partnership and further that there will be no further call of capital.

The last mentioned is an important technique because it is in the nature of a limited partnership that, on the one hand, general partners may share profits and losses but are not obliged to contribute to capital, but on the other, limited partners may also share profits and losses but must contribute to capital. The essential thing, accordingly, is that there must be this initial contribution to capital by the limited partners at large, albeit nominal, or for loans to be made, for the edifice to be structurally sound.

Limited partnerships, being also registrable at the Companies Registry, there is some, albeit extremely tenuous, comparison to be made between partnership shares and company shares in that a share in the partnership can be assigned so long as it is registered and indeed notified in the *London Gazette*. Alternatively, at the outset a partner's capital can be made expressly assignable in any event (capital is otherwise assignable under Section 31 Partnership Act 1890) or not as the case may be, if that is what the partnership agreement prescribes.

Limited partnerships are essentially tax transparent, including when being used as an investment vehicle, being one of a small clutch of tax transparent vehicles so ordained by Statutory Instrument 266 of 1988 (see also Inland Revenue Statement of Practice D12). They are also constrained in terms of promotion and regulation by Section 235 Financial Services and Markets Act 2000 (in succession to Section 75 Financial Services Act 1986), the unfortunate effect of which is that investment limited partnerships should be considered as collective investment schemes, and the party running the project must be authorised to carry on investment business. There are attendant practical and legal issues beyond the scope of this book. If there is the slightest doubt as to whether one or more of the parties carries on investment business, it may be safer to use a company as the vehicle.

## Limited liability partnerships

This book is timely in terms of development and the law as, since 6 April 2001, it has become possible to incorporate limited liability partnerships in England, Wales and Scotland (but not in Northern Ireland) under the Limited Liability Partnership Act 2000. Note 'incorporate'. Limited liability partnerships (LLPs) have separate legal identity. They are clothed with some of the fabric worn by limited companies whilst maintaining certain of the traditions of formal partnership. They retreat in some considerable measure, however, from the difficulties of carrying on business with wholly limited liability.

LLPs have, therefore, been welcomed by professional firms a number of whom have already reconstituted themselves in that form. However, as well as being a vehicle for professional partnerships, LLPs have potential for business ventures at large, including property joint ventures. It is to be expected, therefore, that in time LLPs will become the employers of contractors and construction professionals, and enter into other development commitments as would, say, a company created as a single-purpose vehicle for a developer or itself the creature of a joint venture.

It is therefore necessary to understand the nature of an LLP and the extent of recourse to it and, for that, to attempt some assessment of the economic consequences of dealing with it. When one is dealing with joint-venture vehicles, of whatever kind, the pockets may be deep or they may be shallow, most often the latter for that is part of the rationale for the structure, thus underlining the need to be circumspect and to seek guarantees from deeper pockets where possible. But the wisdom of dealing with any particular vehicle should also be measured by the availability of recourse.

For those in the professions, LLPs are seen as a welcome relief from the traditional partnership with unlimited liability. Larger law firms for example, with the capacity and resources to undertake major corporate finance and other commercial transactions have faced soaring insurance premiums and the prospect of utter ruin if a transaction perhaps running into billions should go off the rails. The pressure for relief has been there for years. Other professions, sometimes working on a smaller scale, have allowed their members to incorporate with limited liability. A mixture of professional reputation and the maintenance of suitable professional indemnity cover may go some way in providing the client with confidence, although it is not the entire answer. Indeed, a company with limited assets may be only as good as its ability and memory to keep up payments of its insurance premiums. There is, however, an emerging recognition that there is only so much responsibility that, ultimately, one can be expected bear. The 1990s in particular also highlighted the general weaknesses in commercial strength and the fall out perhaps showed that there was an enormous disparity between different kinds of commercial venture, particularly so far as the provision of professional services is concerned.

Accordingly, the arrival of the limited liability partnership will be of interest, in its own right, to professional firms whose governing bodies are prepared to allow their use. Moreover, the LLP has clear potential as a joint-venture medium, including for investment purposes (as to the latter, except in particular circumstances). So, if one is a construction professional taking an appointment from an LLP, precisely what is one dealing with, and what is one's redress if the LLP defaults?

First and foremost, an LLP is not a true partnership. Indeed, it is character-ised by being a legal person in its own right, just like a company. In other words it is a body corporate and thus, like a company, can own its own property and incur liabilities entirely separately from the members. Remember that under a true partnership (whether constituted expressly or being a partnership in substance notwithstanding), the members of the partnership can individually bind the partnership. With an LLP, a member binds the LLP instead, as distinct from its members. Tortious default (e.g. negligence) is thus default on the part of the LLP, essentially no differently from a company. Whatever the liability, the members themselves are not individually liable, except in their own right outside of the LLP structure.

To every rule, of course, there is thus an exception and lawyers will point to earlier law to show that, for example, just as directors of a company can be held personally liable for their own negligence, so may also members of an LLP, particularly if the claimant/client can show that he was relying on the individual. Conversely, there are assumptions to be made, much as in the same way as with directors of a company, that members of an LLP are authorised to act, which points up the need for one to be clear that a person is actually a member of the LLP.

Other corporate characteristics are, therefore, the need for registration at Companies House and the filing of accounts there. By contrast, a true partnership's accounts are essentially a private matter. Again, unlike a true partnership, but entirely like a company, an LLP can create floating charges. The term 'floating charge' may leave the non-lawyer cold, but it has signi-ficance in terms of the extent of security afforded by it. For a detailed analysis of different kinds of security, the manner in which security is given and the priority of interests of the respective holders of more than one security is best left to a specialist work. Suffice it to say for the purpose of this book that a floating charge hangs like the Sword of Damocles over the class or classes of assets, or indeed perhaps the whole of the assets, so far as therein prescribed, charged by the floating charge. The LLP or the company having given the charge can deal freely with those assets until the charge 'crystallises'. By crystallisation is meant a demand made under the security, e.g. repayment date has arisen, or an event default such as insolvency. When the security crystallises, it becomes fixed, essentially as if a fixed security had been given at that precise point. Thus, if one is dealing with a company or an LLP and, for example, one is buying an asset, the very minimum act of due diligence is to see what charges have been registered, and if there is a floating charge then to seek a certificate of non-crystallisation. Such a certificate can be given, for example, by a director of the company or members of the LLP. Far better, however, if the chargee itself provides the certificate.

As to insolvency, in broad terms LLPs are subject to very similar insolvency rules as for companies albeit with one important distinction. Sums withdrawn by members of the LLP up to two years before winding up can be reclaimed or clawed back if the person so withdrawing knew or ought to have known that there was no reasonable prospect that the LLP would avoid an insolvent liquidation. It is therefore incumbent on members of an LLP to be alive to its financial circumstances.

The last point also has significance in the fact that, unlike a company limited by shares, an LLP has no capital as such. It may have assets but the legal entity, the LLP itself, is only indebted to its members to the extent of its otherwise unencumbered assets. So, in the absence for example of shares or classes of shares, the members of an LLP can expressly agree what their share of profits may be (and unlike a shareholders agreement so far as affecting all of a company's members, an LLP agreement is not registrable at Companies House).

Accordingly, an LLP is seemingly a misnomer because essentially it has unlimited capacity. Moreover, unlike a company, it is not tainted with the doctrine of *ultra vires* and can do essentially as it pleases. Like a company, it must be incorporated but, like a partnership also, it must be incorporated by two or more persons 'associated with carrying on a lawful business with a view to profit'. Carrying on business 'with a view to profit' is, of course, the principal characteristic of a true partnership. Upon compliance with the requirements of the act as to incorporation, the Registrar of Companies, however, then registers the LLP and issues a certificate of incorporation. Thereafter, changes in members must be notified to the Registrar of Companies. It is the members of the LLP however, who are themselves responsible for filing the annual return and so on, albeit itself very much like a company.

Because an LLP has no memorandum of association or articles of association, and neither does it have what may be described as the fall back provisions of the Partnership Act 1890, it follows that there must be a members agreement dealing comprehensively with the entirety of the LLP's affairs. Compliance with statutory requirements for incorporation apart, it is essential that those wishing to incorporate as an LLP take legal advice. Indeed, even where there is a true partnership, a formal agreement has always been desirable to avoid the imposition of the Partnership Act 1890 where it might otherwise not be avoided.

It is beyond the scope of this book to deal in any detail with the taxation consequences of one kind of structure or another. However, very broadly, unlike companies, but like partnerships, LLPs are essentially tax transparent so that the taxable profits of members are essentially the same as those of members of a partnership. Trading LLPs do suffer certain restrictions on loss and interest relief. Assets held by the LLP will broadly be treated for capital gains purposes as if they were partnerships assets.

An LLP, a true partnership and a company respectively are all identifiable entities for VAT purposes, however, and where an existing partnership becomes an LLP the inference is that a new registration will be required but it may be relieved from VAT on transfer to the new entity under the rules for transfer as a going concern. VAT group rules may also apply.

Finally, therefore, what can an LLP be used for? Amongst other things, it clearly has its prospective uses as a joint venture vehicle. Unlike a limited partnership, there is no need for general partners and limited partners as the entity itself has liability, and as a legal entity it escapes characteristics such as a collective investment scheme. An LLP can also be used as an investment vehicle but it is essential that, as with all joint-venture structures, tax advice is taken as a matter of course. Particularly institutions using an LLP as an investment vehicle cannot take advantage of exempt status, and for that reason LLPs may be seen as inappropriate vehicles for institutional investment.

It cannot be over-emphasised that, however any kind of joint venture is structured, taxation is of primary importance. Moreover, taxation is of no less importance when principals in the development process are dealing with each other on an arms-length basis. Whatever the ingenuity in devising a development structure to accommodate the needs of the respective players, the potential tax consequences for one or more of them may ultimately dictate how the structure will ultimately emerge, or at least limit choices.

## Companies

Companies have been left to the very end and, again, it is beyond the scope of this book to provide even the most condensed source of reference for company law. Companies, however, are to be found trading in all aspects of business, and some understanding of how they work in relation to development, and those who are engaged in it, may be helpful. Many professional firms have themselves become incorporated with limited liability, thus presenting a legal entity which carries on the trade.

First, a company is a 'legal person'. It will thus be the contracting party. The members of the company, however, as shareholders, are thus separate and distinct from the legal entity and so have the protection, so far as it goes, of the 'veil of incorporation'. A company may be limited by shares or by guarantee, or it may be unlimited (in which case its members do, actually, have unlimited liability for the debts of the company). For practical purposes, unlimited companies must be seen as rarely, if at all, having a role in mainstream development activity. For practical purposes also, therefore, let us concentrate on companies limited by shares. The liability of shareholders for the debts and other liabilities of a limited liability company is essentially limited to the amount unpaid on their share capital. For the most part, companies will be found to have their share capital fully paid up. Share capital is, accordingly, a debt of the company to its shareholders and so appears as a debt in the balance sheet, but once a shareholder has subscribed to and paid for his share capital it becomes essentially notional in his hands and the true value is whatever it may be, day to day. This is because, in a winding up, what is left to shareholders is only what is left over after all debts and other liabilities have been met. This may, accordingly, be more or less than what was originally subscribed for. Members' income is represented by dividends, which essentially attract income tax. Growth in capital, reflected in disposal of shares is represented by capital gains. Shares are essentially tradable albeit subject to statute, the articles of association and, if relevant, any shareholders' agreement. Companies are principally governed by a body of law contained in the Companies Acts. If a company is incorporated with limited liability, its memorandum of association defines its objects, and its articles of association govern the relationship of the shareholders to the company and the general conduct of business.

Accordingly, companies are tainted with the doctrine of *ultra vires* and although there are certain provisions in the Companies Acts apparently relieving persons dealing with the company of the burden of verifying directors' decisions, it is unwise to place entire reliance. Just as care must be taken in the formulation of a memorandum of association to ensure that the principal objects, e.g. dealing in property, construction, giving security, letting and so on are fully covered, legal advisers in turn will be concerned to ensure that those with whom a player in the development process is dealing are empowered so to do.

Companies limited by shares are said to be either 'public' or 'private'. Until the Companies Act 1980, upon incorporation a company was essentially a private company which had as its essential characteristic that its shares were not freely tradable in the open market. A company had to 'go public' for that to be achieved. In 1980, the process was essentially reversed so that steps had to be taken to 'go private'. The first myth to be shattered, therefore, is that there is some magic in the initials 'PLC'. There is none. A company may have been admitted to main listing on the stock exchange, or perhaps even to the AIM market, but the underlying notion of a 'PLC' is simply that of a company limited by shares which prospectively, are tradable.

The strength of a company therefore lies, not in its status, but in its assets. But even if it has substantial assets, its skills and effectiveness are another matter again. Chances are that in the context of a joint venture, a company being used as an interface for parties to the joint venture with other players will be a private company, and likely also a nominal entity. Its members may come from a variety of sources seeking, on the one hand, to provide input of some kind to the joint venture and, on the other, to profit from it in some way. Even local government can participate in companies, but only to a very limited extent, essentially to not more than 20 per cent of capital (Local Authorities (Companies) Order 1995), but local authorities have statutory difficulty in making and being seen to make profits, and will thus not be found as parties to joint ventures at large. A company is taxable on its own profits and on its capital gains, through the medium of corporation tax. There is no inherent tax transparency. Dividends payable to its members are taxed as income in their hands. However, in the context of property joint ventures, or indeed any other kind of business joint venture, the essential characteristics of a limited company stand behind a structured arrangement between its members, or participators as they are known. In the context of property joint ventures, the roles of the participators in the joint venture will thus be reflected in a joint venture agreement. Such an agreement will prescribe precisely what each participator is to do, whether to lend cash (to the joint venture company) to provide skills (including professional services) or to make assets available (e.g. land) and so on. One may therefore expect participators both to be members of (and thus potentially to profit from) the joint venture company and also to contract with it as a separate entity in other ways. Note, for the first time the term 'joint venture company' has been used. It has no legal significance at all and is at best descriptive: similarly also the 'regeneration company', a term now beloved of the regeneration 'industry'.

The relationship of a company to its participators as members should therefore be seen as standing behind the company's relationship with the participators as separate contracting parties to the joint venture, or other aspects of the development, to the point where the relationship may be seen to override the articles of association and thus perhaps need itself to be notified to the Registrar of Companies, as adding to or modifying the articles of association. The business of the joint venture apart, the relationship of the participators may thus have to be set down in a shareholders' agreement, and whether the arrangement is seen as requiring registration is a matter for legal advice as much as preparation of the agreement itself. In any event, no joint venture should be attempted without taking legal advice combined with tax advice as appropriate.

The overriding of corporate structures and benefits may manifest itself in a number of ways. Participators, in contracting separately with the company, may draw their first tranche, as it were, of benefit or profit. For example, one of the reasons for entering into a joint venture may be to defer the expense to the developer of acquiring land. Carrying costs are all in development, and the less that has to be borrowed the better. Thus, a developer and a land owner might reach agreement that they will create a development company, and the land owner will sell the land to the company but will defer the price or 'consideration'. This may likely be a purely contractual matter, or perhaps through a declaration of trust for the joint venture company (which, incidentally will carry stamp duty land tax), or by transferring as such to the company while providing a loan, and so on. There will, incidentally, be little point in securing that loan by primary security on property as, for the purpose of raising development finance, the arrangement will be seen as unbankable. A lending bank must have primary security and so deferring consideration without impairing security will likely be the general aim.

Again, as often happens, a developer may have skills but limited capital resources, but its chosen contractor may have skills as a contractor, and not as a developer, but also have financial resources. The contractor may, therefore, agree to 'lend its covenant', by which is meant that it is prepared, under the terms of the joint venture, to guarantee financial commitments. Whatever the arrangements, the implication is that in return for providing whatever input, participators may have independent contractual arrangements which afford them priority returns, which means that the value of their shareholdings as such may be of little or no consequence.

At a slightly deeper level, as with all commercial arrangements, one has to consider fallbacks in case of default. It is not enough simply to consider the matter in terms of legal liability, the loss or damage suffered and the result of damages. Who will pay and will there be sufficient assets? Thus woven into the fabric of any corporate joint venture or shareholders' agreement should be, so far as possible, events of default and the consequences of default. For example, in relation to the particular function which a participator performs, he may be required in case of default to forego the whole or part of the benefit which he expected to derive. His shares, if potentially valuable according to the chosen structure, may also become forfeit and divisible between other players. More particularly, he may be deprived of dividends, contrary to the status of his shares, apart from his other contractual reward, and so it will be seen that arrangements such as these, may, in the event, override the provisions for payment of dividends contained in the articles of association. It is all a matter of sticks and carrots according to the business objectives of the participators and all must be contractually catered for. Creative business is reflected in and governed by the corporate structures and contractual arrangements governing the business of the company.

Occasionally, a joint venture company can be so structured as to be tax transparent. The starting point is that a company, as such, is not tax transparent at all. It is a legal person and a taxable entity. Such tax transparency as there can be must thus be entirely contrived, and thus requires careful legal drafting. However, the flow of benefits can be so weighted that the company is wholly nominal and the interests of the participators simply flow through it contractually, as it were, 'without touching the sides'.

So how should those dealing with any company react? As a starting point, assume nothing. The company may be so structured as to have no substance or value and be no more than a nominal entity created as an interface for dealings. That, unfortunately, is prospectively the employer of the professional team and contractor. The fact that the participators in the company may themselves be of substance is potentially of no value at all, even in a winding up. Granted, in an insolvent winding up those who are directors or quasi-directors, that is to say those having a governing influence, may on the application of the liquidator to the court be found to have personal liability, no reliance whatsoever should be placed upon that prospect. It is entirely a wait-and-see situation, the value of which, if any at all, can only be tested at the time. The company may not be a joint venture vehicle at all but have been created solely for the purpose of satisfying a bank which requires to take control of all of its assets, however small, in order to have the ability to appoint an administrative receiver to run its affairs (Insolvency Act 1986 again). That is the least to expect of any well-advised developer. Whatever the commercial pressures on, say, a professional adviser seeking business, he must accept that if he is to be paid at all, then the trust he places in the people with whom he is dealing must therefore be extremely well placed. He must understand that whatever the level of trust, the entity with which he is dealing may not be resourced in the event, and he may receive nothing. Witness the 1990s' recession, when developers collapsed and even where principal companies guaranteed the obligations of their single-purpose vehicle subsidiaries, those guarantees were often useless too. If a professional firm or a contractor is then required to provide a warranty, say to a bank or financial institution, neither does that imply in any respect whatsoever that that bank or institution will be responsible for paying professional fees, unless it is so contractually provided.

There is a simple lesson to be learned from this chapter, know with whom you are dealing. There is no body, even a government agency, which has a bottomless pocket, whatever the legal enforceability. Proving liability in contract before the courts, of whatever kind, and the doing of justice are noble concepts, but empty ones if the assets are not present to meet the liabilities. It may be that *ulta vires* would already have intervened to curtail liability. Even if it has not, and liability is shown, the question is, always, what are the resources. The business of joint ventures is the securing of contractual obligation, incentive to perform and safeguards against non-performance. If the resources available are not known and assessed, if the fall back on default is not provided for, the contract was worthless to start with.

# 3 Development and forward funding agreements

## Or: Now, let's really party!

Unless one is dealing with a developer, fully resourced and having no dependence on joint venture partners, banks, institutions or tenants, it should be assumed that the appointment of any professional adviser to a project will probably not be a free-standing, or at least freely negotiated, contract. The developer will thus have to have regard, if not to actual then to prospective requirements of the performance criteria to be contractually required by each of them. The landowner, funders and others, in turn, will be testing the quality, not only of the developer but particularly of those whom he engages for the construction of the project, and more particularly again the suitability of their respective terms of engagement.

Even if a developer has a close working relationship with a professional firm, the requirements of a particular funder may turn a long-standing relationship upside-down. For example, as will be seen later, the requirements of a bank lender may seem harsh. The reality is that banks are in the business of lending, and lending implies repayment. Repayment, in turn, implies realisation of the asset which, in terms of construction, may yet be incomplete. Engagement of a professional firm, however reputable, might be seen as a close relationship, but in the eyes of a bank it is simply part of the package, and to be treated as such. Later, in Chapter 7 'Construction procurement in the development process', we shall consider some of the influences of other players, including lenders and financial partners, on how terms of engagement may be negotiated and ultimately emerge.

From the first two chapters, it will already be apparent that development is essentially a contractual matter, comprising the interaction between a number of players with different functions. Thus, all development has an element of 'joint venture' in it, albeit a term inherently incapable of precise definition. Nothing can be documented, still less any structure assumed until the players are identified, their precise commercial objectives are set down and how far these should be secured. Layered on to this are issues of taxation and *ultra vires*. There is no order of priority in these factors, they are simply matters which, in any order, have to be addressed and resolved.

However, behind the scenes of every construction project may lie a series of complex contractual arrangements bringing together the principal players.

As between themselves, they need to recognise their respective functions. For example, a local authority landowner, in engaging a developer to carry out a town centre development or some other project in which that authority has a long-term interest, cannot ignore the interests of funders and tenants. It does so at its peril. Either the project will thus be rendered unbankable, or tenants will go elsewhere, and it goes without saying that contractors and construction professionals also want the certainty, of a certain kind, that what they are being asked to do is confined within identifiable parameters and that they will be paid for it.

So much it seems is obvious, perhaps to the most casual observer. To the lawyer involved in the development process and in particular in the negotiation and settlement of development agreements, it is still sometimes as if the respective players in the piece visit Earth only occasionally, and are clearly happier at home in their respective abodes around the universe. It is usually, extremely difficult for contractors and professionals to engage with the process, albeit they must to so in the interests of their business, not least because they are usually excluded and then presented with a *fait accompli*. A development agreement that fails to engage with the expectations of construction contracts may itself be flawed, perhaps with detriment to the project overall.

This chapter, therefore, addresses the basics of development agreements, and in particular two major development media, the classic landowner/ developer agreement and the forward funding agreement.

## The classic landowner and developer agreement

The landowner does not need to be a local authority, indeed such an agreement may be entered into between any kind of landowner and a developer (and the developer itself may be the product of a joint venture, or be a nominal entity created for protection from insolvency consequences). That it should be entered into at all implies that the landowner has some valuable interest which will survive disposal of the land to the developer, and which is directly related in some way to the development itself.

If the landowner is simply interested in the financial outcome, then some kind of profit share arrangement, or the imposition of some limitation on use which can be lifted or modified in return for payment, may be more appropriate instead. It is all, lawyers say, a 'matter of degree', but a material interest in the built outcome implies the need for a development agreement. If the developer is being engaged merely in a project management role, then that is the kind of agreement he may expect to receive (and for that service he may therefore charge VAT).

A landowner/developer development agreement, however, implies that the developer is to receive a major interest in land. Moreover, insofar as the landowner desires to regulate the ongoing use of that land, once the development has been carried out, the implication is that the interest which the developer may acquire is likely to be a lease. Leases are commonly granted for anything up to 999 years, albeit that the expectations of municipal practice may be somewhere between 125 and 250 years. Value depends on the nature of the project and the landowner's perception of his reversion, concerning all of which the reader should turn to an investment surveyor. The longer the term, the lease may ultimately be regarded, in layman's terms as a 'virtual freehold'. From a valuer's point of view, a lease may also be so seen. The value of a lease in the hands of a developer/purchaser must, therefore, be seen by contrast in terms of the value of the reversion to that lease in the hands of the landowner.

Accordingly, taking the example of a lease for several hundred years, the remoteness of the reversion is but one factor in determining the value of the lease to the landowner/reversioner. If the rent is entirely nominal, a peppercorn rent as it is known, there is clearly no financial value to the reversioner unless the term is short and, once the lease is up and running, the reason for having a lease at all may run in other directions, such as control over use or management alone. The reason for having a lease is to regulate the continuing use of the land according to the intended benefit to the landowner, albeit the nature of that regulation may itself have an effect on the relative values of the lease term and the reversion, such regulation being secured not only contractually in covenant, but also in the ultimate sanction of forfeiture.

Thus, taking the example (but only one kind of example) of a town centre development, a local authority may wish to see an area of land rebuilt in a certain way, perhaps providing facilities for community use whilst affording the developer the ability to create commercial space, so that it becomes a capital asset in the hands of the developer and from which it may derive income, that is to say, rent. The local authority's interest may lie in regulating the use of the development and/or perhaps deriving a proportion of income or whatever other interest the local authority has in the long term.

The lease, accordingly, is the long-term instrument for the relationship between the landowner/local authority and the developer/successor. The development agreement, meanwhile, is the engine for creation of the entity, the development that forms the core subject matter of the local authority's interest for the future.

There are, thus, two entirely separate documents in the structure, performing their respective functions for the immediate and longer terms respectively. Granted that there is such a thing as a 'building lease', modern development and construction requirements inevitably mean that to follow this route would leave the lease burdened for ever by hand baggage that had become redundant over time, and within legal limitation periods mostly in the very short term only. A modern lease should be seen to stand alone.

The developer may be required to buy his lease of the land. He may, instead or as part of the consideration paid for the land, be required to create further development if not within the development site then upon land retained by the local authority. There is no limit to what this might consist of, whether it be highway works or the creation of some facility is not an issue. What is important is the developer's obligation to provide whatever it is that the local authority requires, either partially or wholly in lieu of consideration, that is to say, the price to be paid by the developer for the land which he is to develop for himself. It may be, however, be that the landowner in question, local authority or otherwise, is simply interested in a share in development profit. This can be derived in a number of ways without need for a development agreement of the kind contemplated by this chapter including, if a slice of rental income is not in prospect, a slice of proceeds. This implies instead a profit-share arrangement under which development costs, defined within close parameters, are set against disposal proceeds (deemed or actual), again defined within close parameters. Whatever the actual cost of the development to the developer, what is to be allowed as the development cost for the purposes of the calculation will be set against the outcome of the (deemed) disposal. This further implies that disposal will be on strictly controlled terms or, in so far as disposal itself is not controlled, then upon notional criteria (deemed disposal). There is no rule of law which says that development agreements of any kind should follow this or that pattern: development agreements are matters of contract, and the formulae within them result from commercial negotiation, albeit tempered by legal and taxation constraints.

If the landowner's interest includes a proportion of prospective occupational rents, the financial criteria have to be secured, and a leasehold structure achieves this. If not under the enabling agreement (assuming the lease has not yet been granted) then under the terms of the lease itself, disposals to tenants can thus also be controlled. The essence of a lease, hence the reason for using a leasehold structure at all, is that breach of its terms can ultimately lead to forfeiture. A lease, therefore, combines within its terms the means of controlling dealings to provide ultimate security for the financial objectives. Clauses concerning dealings with leasehold property are known as 'alienation' clauses. A prospective occupational tenant of the development will be no less concerned that the terms upon which he takes his own lease are in accordance with those provisions, otherwise his lease may be undone by the courts, or dealings with his own lease may be unduly restricted.

The preliminaries aside, what does a landowner/developer agreement contain? It is clear, first, that the developer has to provide something to the landowner, simply for the privilege of acquiring the land in due course. Second, he must meet the landowner's requirements as to carrying out the development whether on that land or on retained land. Thereafter, the developer is confined by whatever terms the resultant lease he takes imposes on him.

For those who are concerned with town and country planning, there is an obvious distinction to be drawn between what is freely contracted on the one hand and what a planning authority, in that capacity, may impose on the other. It is well established, in general planning terms, that a local planning authority cannot in that capacity impose conditions affecting land beyond the planning site. It may, however, impose as a condition of granting planning permission a requirement that the applicant enters into a statutory agreement concerning, so far as may be, relevant works on that other land, for example highway works. A development agreement, by contrast, is a freely negotiated contractual instrument. If, as a consideration for affording the developer the opportunity to develop a particular site, the landowner wants works carried out elsewhere, so be it. That is his contractual freedom.

Before we consider the contents of the development agreement, one further point should be noted. The development agreement may prescribe that all or part of the works should be carried out, in accordance with its terms, before the major interest (assumed lease) is granted. Assuming the agreement is properly drawn the agreement itself can be secured by appropriate charging instruments to a bank lender. As chargee, the lender will, in turn, be entitled to call down the lease to the developer, so long as further security is then taken upon the lease for continued protection, but will be empowered (see also below) to assign (transfer) the agreement to a third party in order to realise its security, or call down the lease to its nominee so far as the development agreement permits. Banks are not in the development business, they are lenders, not developers, and the basis of making loans is security.

Perhaps the most often fought over provisions of a development agreement are thus those relating to a bank's security. The issues are, actually, quite straightforward. The landowner has negotiated with a particular developer and has decided that he has the confidence in that developer to deliver the desired development and bring about the fulfilment of the landowner's objectives. The bank, by contrast, exists to lend money in order to produce a return for its depositors and, on the way, to take some profit for itself. The banker's business will collapse unless the depositors' savings are safe, or at least as safe as can be. In crude terms, that is the essence of the relationship with the developer. Thus, if the relationship should fail, the bank must be able to recoup its losses so far as may be. It could appoint a receiver to build out the development, where there is no separate landowner's interest to accommodate, and proceed with implementation of the development agreement. Its judgement may be, particularly early in a development, to transfer the benefit of the development agreement to another developer, who may again become secured to a banker and so on. The developer's bank meanwhile achieves (hopefully entire) repayment, but not necessarily. However, it is in the nature of development agreements that developer insolvency is usually also expressed as an event of default so, unless the landowner is willing to allow a receiver to operate the agreement, disposal is the lender's only practical option.

Needless to say, the landowner has some interest in all of this, not least in the insolvency of the original developer. It follows that the development agreement should contain appropriate provisions contemplating its use for bank finance, including criteria as to the quality of the lender itself. It should, however, then go on to deal with questions of disposal and insolvency. A chosen developer is unlikely to be allowed freely to assign/transfer the agreement, because of the landowner's need to ensure that the developer's original obligations should be carried out by him in particular. After all, the original contract should be seen as a fairly close arrangement. However, the landowner must recognise the position of the banker who may, *in extremis*, be compelled to look to the wider market. This implies that, in case of developer default, a new developer will have to be found. Accordingly, where the banker must dispose, then the landowner must anticipate having to acquiesce in acceptance of a new developer. All of this must be achieved contractually through criteria carefully laid down in the development agreement, in order to retain sufficient quality in the developer's covenant.

It will, therefore, be immediately apparent that in entering into a development agreement, a landowner must cater for two sets of circumstances. The first is the primary obligations on the part of the developer and the performance criteria associated with them. The second is the basis upon which the developer may finance the project and how to deal with the needs of any funder. Failure to accommodate the latter renders the project inherently unbankable.

The whole fabric of a development agreement will be of interest to bankers, purchasers, investors, tenants and so on, including not least those engaged in actual construction. Whatever the interests of the landowner, conflicting interests of those who are looking to benefit from the development must be taken into account. Otherwise the landowner's ambitions will be fruitless. Bankability is the key. Moreover, even if the development requirements themselves are seen as right and proper, the denial to the banker of the opportunity to remedy any default of the developer or the imposition of subjective tests on performance criteria may render the agreement no less unbankable than if it had contained unreasonable demands at large.

One further factor needs to be highlighted. The agreement can either lead to the grant of a major interest in due course, say upon practical completion of a development, or it can be allied to an immediate grant. There is some law, and perhaps a rather larger measure of myth and sentiment, underlying the debate. An agreement for lease, or indeed any contract underlying an interest in land creates an 'equity'. That is to say, it creates a right to acquire in the circumstances set out in the enabling agreement/contract. An equity is not the same as an estate in land. An 'estate' in land is a legal concept, an equity by contrast, deriving from equitable principles, is not, and remains no more than an 'interest' and the two are distinct.

There are two legal estates in land, the first the 'fee simple absolute in possession' (freehold) and the 'term of years absolute' (lease), both deriving from statute (Law of Property Act 1925) which finally removed the feudal system and, along with a series of other statutes in the same year, finally brought English land law into the twentieth century. These concepts form the backbone of certain kinds of security. One hears, for example, of a 'legal charge' and then of an 'equitable charge'. The two do not, of themselves, simply attach to a legal estate or an equitable interest. The instrument of security, and the entity secured, are separate and distinct notions. Thus, one can create an equitable charge over a legal estate. It gives no power to exercise a right of sale, for example, but creates a priority interest (in the case of registered land) which must first be satisfied. By contrast it is feasible, technically, to create a legal charge over an equitable interest, albeit to limited effect. It affords a power of sale, in theory, but the equity secured thereby is an inferior interest hedged around by the terms upon which it is dependent for enlargement into an estate. (A legal estate is registrable as such at HM Land Registry, and the resulting title certificate constitutes the legal title. An equitable interest enjoys no such status and may at best only be registered itself against the relevant legal title.) Thus, for the banker, there is first the issue of enforcement of the security, and then, on top of that, enforcement of the equity against the owner of the legal title from which the equity is derived. (Other kinds of instruments, simply contracts other than in relation to land for example, may only be secured by assignment as such, coupled with a 'right of redemption'.) These niceties are, it is clear, best left to lawyers.

Part of the fallout from these conflicting interests is the growing practice of granting the major interest at the outset, alongside the enabling develop-ment agreement. This may seem self-defeating, in that the landowner has seemingly thrown away the carrot and stick, but even the lawyers have thought about that. The grant of the lease, at any time, creates an estate in land. It is therefore susceptible to full legal security and provides the best legal protection for a bank or lender providing development finance. To remain effective, the development agreement and an early grant must, therefore, be contractually allied. From the lawyer's perspective, it is better that the lease is silent in all respects as to its enabling agreement, at least so far as may be, so that when the development obligations of the agreement eventually fall away the lease stands alone without the need for referral back to a preceding document. The development agreement will thus be expressed, in so many words, to be supplemental to the lease so that the full raft of default measures relating to the development, contained in the agreement, can be brought to bear upon the lease as if they formed part of the lease itself. In that way, whilst the lease is clothed with the quality of primary security, the development agreement prevails, including the ultimate sanction of forfeiture of the lease, even though the substantive breach may lie in the development agreement. So, to the development agreement itself, what it contains, and how the relationship between developer and his professional team and contractor are shaped by that contractual relationship. That it is further shaped by the developer's relationship with his banker follows later.

## Terms of the development agreement

There is a certain pitter-patter to development agreements, and some or all of their provisions frequently find themselves being incorporated into building contract requirements. They are then therein expressed to override the detail of the building contract, thus presenting a perhaps impossible task for the contractor, the contract administrator and the professional team. From behind the lawyer's desk, one sometimes wonders how much notice is taken of the detail, how far it is understood and whether advice was taken by the contractor before incorporation into the employer's requirements. For the most part, it is only the operative provisions relating to the procuration and construction of the development that are relevant. The remaining provisions may ultimately be no less relevant, particularly in the case of default and where those involved in the construction may find that the introduction of new masters or, worse, collapse of the entire project and the realisation of fees, may lead directly back to its provisions.

### Recitals and definitions

A lawyer skates over these at his peril. They not only set the scene and the context, but the definitions are critical to interpretation. Neither is the use of capital letters an idle throw back to times when the capital letter gave importance to a word. If a definition has a capital letter, it is the definition to which one looks when that word is used with a capital letter. If the same word is then used in lower case, its ordinary meaning should usually be attributed to it instead.

Immediately after these provisions there may often be an interpretation clause including provisions such as an assumption that reference to a statute includes reference to a statutory modification. The absence of such a provision has the unfortunate side effect that if, for example, a clause in the agreement says that one must do this or that in accordance with a certain statute, and that statute has since been modified, the provisions of the original statute will remain contractually incorporated. Sometimes, of course, this may have been the intention at the outset.

### Grant of lease and other conveyancing issues

These provisions are perhaps of less interest in the construction context. These provisions may succeed the development obligations but people's differing drafting techniques mean that there is no inevitable and usual order of material. The development obligations in particular may be hived down to a schedule or schedules. For a schedule to operate, there will be a related provision in the main body of the text saying, for example, that the developer shall observe and perform the obligations set out in that schedule. This author's preferred layout (as found in his contribution to volume 38(2) of Butterworths' *Encyclopaedia of Forms and Precedents, 2000*) is to schedule the development obligations. Other forms of development agreement, including for the provision of development finance, are often similarly ordered for consistency. Another approach to be found is to go into development obligations almost straightaway leaving conveyancing issues, alienation, default and disputes and so on to the end of the document. A variety of things may yet find their way into schedules however.

## *Alienation*

This clause goes to the root of bankability and to the relationship of the developer with the landowner. (A contractor or professional required to have regard to the development agreement should read this provision too. It shows who the new masters might be in case of employer default, and operation of novation provisions in a warranty, for example.) Its operation, particularly in case of default, and the combined effect of the alienation and default clauses, thus prospectively impacts on the relationship between the developer and all others involved in the process including, very particularly, those involved in construction. There is a direct correlation between alienability of the agreement and the banker's expectations in construction warranties.

First, the chances are that the developer will not have been chosen casually. The inference is that if the landowner has been careful, the developer's resources and skills will have been the subject of some due diligence, with the developer perhaps being in competition for the project. The agreement will not, therefore, be readily alienable, i.e. the developer cannot simply transfer it to someone else and exit from the project. Similarly, if the lease has been granted at the outset, operative provisions in the agreement will ensure that whatever the terms of the lease, during the development period the provisions of the agreement relating to alienation will override.

However, the agreement cannot be entirely personal to the developer if the developer also needs to raise finance (and the directions in which the benefit of the agreement may go will, in turn, be later reflected in the multiplicity of construction warranties). First, the agreement (and the lease, if granted) must be capable of being used as security. Commonly, it is provided that the agreement may be assigned/charged to a lender first approved by the landowner, whose approval shall not be unreasonably withheld, for the sole purpose of providing finance for the development. Second, upon realising its security the lender must be capable of assigning on to a new developer. It follows that that new developer will also require to be approved, and the arrangement will be unbankable if the criteria for approval are less than objective. The bank's receiver, if any, will not usually operate the agreement in case of developer insolvency, indeed such an event will usually be expressed as an event of default under the development agreement (see landowner and developer agreement above), but a prudent banker might consider that if the development is almost done, a receiver has a role to play and may seek an amendment along those lines.

Alternatively, if institutional finance is involved – as under forward funding next described in this chapter – the agreement (and lease, if relevant) must be capable of assignment to that institution. Again, the institution must meet appropriate criteria for acceptability. In this case, however, and as will be discussed more fully later, the institution or fund displaces the developer entirely, but then sub-contracts the development obligations back to the developer. Unlike a bank which needs to facilitate repayment of its loan, a funding institution can expect rather less sympathy having stood in place of the developer and during the development period will be unlikely to alienate the agreement further. After all, its remedy then is to sub-contract the work to a new developer or carry out the development itself. Once the development is carried out, it can deal with the lease in accordance with its terms. So what happens in case of default?

### Default

Those dealing with the developer, not least those concerned with the construction itself, have a direct interest in the consequences of default. The landowner has further things to think about if loan finance is employed. Default may arise under the development agreement, whose terms will have been negotiated between the landowner and the developer, but default may also arise under the loan finance agreement. In other words, an issue may arise between bank lender and developer which perhaps does not affect in any way the implementation of the development agreement. One thing is certain, however, which is that unremedied default, whether under the development agreement or under the loan finance agreement, will likely lead to a fundamental shift in the relationship between the developer, its contractor and professionals. It may lead to the installation of new masters or, by no means impossibly, collapse of the entire project or at least the construction-related elements through the introduction of an entirely new team.

Events of default under the development agreement will include, of course, insolvency and, inevitably, the prospect of irremediable default in the development obligations. The development obligations will be hedged around with a series of checks and balances, approvals and so on, to ensure that performance is delivered. Inevitably, timing will also feature amongst events of default, whether the development or a stage of it is not achieved within a certain timescale.

The consequences of the foregoing are that the right to carry on the development may determine, and it is for the developer's lawyer to ensure that the agreement contains sufficient safeguard for due notice to be given and for the opportunity of remediation. Otherwise, the consequence for the developer may be, ultimately, that everything is forfeit to the landowner including what has been built so far, with the added consequence that the building contract, professional appointments and so on will become incapable of performance. (The legal entitlement on the part of the contractor of the professional team to recover their losses from the apparent repudiation of the building contract and related appointments will be of no value without assets in the hands of the employer to meet those liabilities).

In order to ensure that advances made under its loan facility to the developer are not sacrificed, a lender will thus be concerned that the development agreement contains appropriate fallback provisions. These should include the opportunity itself to remedy (or rather to procure remedy of) the default, and to that end to be entitled to be notified at the same time as the developer of any default. Statute provides in case of leases (and often the provisions are imported expressly into development agreements) that forfeiture for breach of covenant cannot operate except upon service of notice under Section 146 of the Law of Property Act 1925 which, in simple terms, requires determination by the court so affording the prospect of exercise of the court's equitable jurisdiction to grant relief from forfeiture.

It follows from the foregoing that, to be bankable, the agreement must also contemplate the extreme circumstance that the lender must have the ability to put in place a new developer. As mentioned before, banks are lenders, and not developers, and so the appropriate medium for continuing to operate the agreement will be a new developer, rarely receivership (but see Chapter 4, 'Forward sale, loan and occupational tenant agreements').

In Chapter 4 we shall look briefly at development loan agreements. They rarely rise to the surface in construction terms, but those who are engaged by the construction process in any part of the development may notwithstanding feel the effect. Prescribed events of default in loan agreements, entitling the lender to realise its security, will inevitably include default under the principal development agreement. They will then go on to deal specifically with events of default arising under the loan agreement including, in common with the principal agreement, insolvency. Breach of the loan agreement, in circumstances wholly unconnected to the principal agreement, may be rare in practice but the possibility must be catered for. In that event, and if it does not trigger insolvency which is already catered for under the provisions of the principal agreement, the implication is that the lender may have to exercise its power of sale. The lender, in exercising its power of sale, now has to look to the terms of the development agreement to see in whose favour it may be alienated (transferred) and so once again the provisions of that agreement are in issue. The only practical way out is through objective criteria for alienation being incorporated in the development agreement at the outset. The reaction of the banking fraternity is entirely predictable and a lawyer's failure to recognise this when negotiating the development documentation will have inevitable consequences.

For those involved in construction, some of the fog may now begin to lift over why construction contracts are so framed (or standard printed forms are so amended), e.g. that the benefit of the building contract may be freely assigned to a mortgagee, it being in the nature of such a contract (and also a professional appointment) that the only way to secure the contract as part of the mortgaged assets is to assign it, coupled with a right of redemption (contrast a mortgage or charge over land) and these will be discussed later in Chapter 4. There may be new masters or there may not. If not, legal remedies against the employer may or may not bring their reward. It all comes down to resources. With luck, a lender or an institution anxious to have a loan repaid or to ensure that its investment is not wasted, will be able to reconstitute the development. If there is a new developer then possibly the development package may be transferred, and the team retained. Assignment of a warranty may also be a 'rescue' and it then remains only to ensure that contractual commitments are taken from assignees ('novation'). There is an important but simple point of law which has its application throughout contracts of their many kinds: the benefit of a contract may be assignable, a burden alone is essentially not assignable, but these principles (and their exceptions) can wait until Chapter 7, 'Construction procurement in the development process'.

## Disputes

Disputes under development agreements should in general terms be distinguished from disputes under construction contracts. First, if the related agreement provides for the transfer of a freehold or grant of a leasehold (of not less than twelve months), provisions which might otherwise be construed as a construction contract under the Housing Grants Construction and Regeneration Act 1996, do not apply. There may be a parallel dispute under a construction contract although, for the most part, the development agreement and any relevant construction contracts will run on entirely separate lines. Exceptionally, however, a dispute between landowner and developer may lapse into default elsewhere having first placed the developer in breach of its construction contracts, e.g. failure to secure approval of a stage of the work which must be done before a further stage can be commenced. For the most part, the nature of disputes between landowner and developer is probably going to be entirely separate although the subject matter may be common and lead to separate disputes under the construction contracts. The absence of compulsory adjudication under the development agreement means that its formal disputes procedure, however that is framed, will be immediately available and will proceed accordingly.

## Development obligations

These are the matters which are the most likely to be reflected in some form or other in the construction contracts for the development, as mentioned above these being sometimes imported wholesale inter the preliminaries. It is, therefore, helpful to understand the mind of the landowner, and the fact that if there is a development agreement at all reflects his own agenda.

For the purpose of this part of the discussion, let us assume certain possibilities, first, that in lieu of part or whole of a monetary consideration for the development site, the developer may be required to carry out works on other land. Again, assume that the landowner has an interest in the developed site, possibly as a occupier of part of it, or because perhaps it has a share in prospective rental income, perhaps it is concerned as a local authority that public rights of access and use and enjoyment are to be made available or, again, the development needs to sit happily with some other activity. Any of these reasons, and no doubt many others, may give rise to a direct interest in the nature of the development and its conduct. The developer will thus not be a free agent and may be subject to a considerable measure of control.

The landowner will require planning and all other relevant consents to be obtained. Consents clauses habitually require, or at least purport to require, that all relevant consents be obtained prior to commencement of the works. As construction and development professionals know only too well, this is an unlikely prospect, particularly in the case of substantial and complex development. Whilst planning permission is an absolute prerequisite, there may be reserved matters or elements of reserved matters which may be in train after construction has been commenced. Further into the development, the supply of pre-manufactured components, say a particular kind of window frame, may dry up, thus necessitating the need for a detailed change so as to admit a new type. Again, approval of plans by the building inspectorate often takes time and so the works will proceed in the meantime in accordance with building regulations. A well-drawn consents clause will recognise the practicalities and insofar as it does not, the landowner has to ask himself whether crying 'Breach!' is a practical proposition.

The approval of the contractor and the professional team, if so prescribed, is not a matter to be taken for granted. Such may be the interest of the landowner in the built outcome that the identity of the contractor or contractors and each and every member of the professional team may be of particular significance. A developer with a long and established relationship with professional firms may find that he is channelled into use of professional firms which the landowner (or its own professional advisers) have worked with or whose professional reputations are well known.

Terms of engagement are similarly also important, as well as the terms of related warranties. The subject will be dealt with more fully in Chapter 7 and the developer has to remember at all times that not only the choice of contractor and professionals but also the form of the building contract and terms of engagement need to be satisfactory to the landowner, to any funder, prospective buyer/investor and, where relevant, occupational tenants. It is these concerns that developers' lawyers have to keep in mind at all times, and which lie behind increasingly lawyer-driven construction documentation.

The contractor and professional team will be working principally to their respective contracts with the developer. The developer, in turn, will be working to the development obligations set down in the development agreement. The less compatible these are, the greater the risk to the developer of precise compliance, with one perhaps comprising precise default under the other. That is an extreme position, but by no means a fanciful one. Take, for example, the process of supervision and approval of stages. The contract administrator under a building contract, for example, has a fair measure of independence. Whilst beholden to his employer, the developer, he must act fairly and be seen to do so. Let us say, therefore, that he has issued a certificate as to completion of a certain stage under the building contract, thus entitling the contractor to a stage payment. The landowner, however, may also be subjecting the developer to a regime of supervision, usually through an approvals procedure conducted by a professional engaged by the landowner for the purpose. The development agreement will usually say that approvals under it shall not be unreasonably withheld. That professional, however, is simply acting as the agent of the landowner and a dispute between landowner and developer is squarely between just the two of them. A disapproval by or on behalf of the landowner, for the purposes of the development agreement, clearly cannot bind the contract administrator but, if the pendulum swings in favour of the landowner, and assuming that the result is a fair one, it certainly brings into question whether the contract administrator's decision was a correct one, and that matter may need to be revisited.

It is generally inappropriate in a development agreement for the landowner to prescribe, say, that the developer shall instruct the contract administrator not to issue a certificate of practical completion unless he shall have notified the landowner's representative so many days in advance, as is sometimes seen. In practical terms, given that the contract administrator is going to be working fairly closely with the landowner's representative, such a procedure may well work in practice. However, a prudent developer, whilst translating relevant provisions of the development agreement into the administrator's appointment, will be wise to ensure that the development agreement is so framed as to make clear that all such provisions, if the landowner insists on them, are expressly without prejudice to the contract administrator's duties so that his independence is not compromised.

Within the development agreement will thus usually be found provisions regarding stages, practical completion, defects liability and so on. All performance criteria should be drawn in such a way as to be subject to objective tests. A subjective test does not merely erode the prospects of satisfactory conclusion of the development agreement. It also means that if an event of default is triggered, the developer may be deprived of the ability to operate his construction contract and professional engagements. In a worst case scenario, when his licence to be on site is terminated through default, his construction-related contracts may collapse. Moreover, if faced with proceedings from his construction team, he may not be able to plead the legal doctrine of frustration of contract if he placed himself in peril through his contractual relationship with the landowner.

It might be different, however, if the construction team were, before being engaged, expressly on notice of these arrangements. Taken to its logical conclusion, however theoretical, the construction team would need to enquire as to every contract which might adversely affect their own. The practicality is that subjective tests by landowners of performance criteria are rarely applied as being inherently unbankable. Insofar as they are applied, there must clearly be a good reason and, in those circumstances, either the project can be done or it cannot. There is, indeed, one very good example which is the matter of timing. Development agreements of their various kinds impose timescales for a variety of reasons. The long-stop date, or the back-stop date as it is sometimes called, usually means precisely that. Such dates are also found in building contracts. The absolute of a long stop should therefore be imposed against the background of compatibility of delay clauses and, above all, that the long-stop date in the development agreement should never be prescribed to fall on a date earlier than the long-stop date in the building contract.

The conclusion to be drawn from all of the foregoing is to be circumspect. It is simply impractical for those involved in the construction process to become involved to any great extent in the relationship between the developer and others upon whom he depends for availability of the site and of his development funding. However, where a development agreement is made an express subject of a construction contract, it should be read with care and legal advice may be necessary.

### Stamp duty land tax

It is timely to mention stamp duty land tax (SDLT) which replaced stamp duty for all land transactions with effect from 1 December 2003. For those whose acquaintance with either tax is limited to the buying of a house, superficially there may seem to be little more than a new label. Nothing could be further from the truth, which is that SDLT is a complex tax derived from swathes of the Finance Act 2003 and a number of weighty statutory instruments. It is early days, but it has the prospect of influencing the way that we structure development transactions and their supporting financial mechanisms, all of which are the life blood of the eventual construction process.

Stamp duty, in relation to interests in land, was essentially a duty on instruments, transfers, leases and the like. Stamp duty was, at one level, a voluntary tax in as much as one was not liable to a penalty in any event if the tax was not paid. However, so soon as one needed to register a deed at the Land Registry (the initial HM being dropped when the Land Registration Act 2002 came into force on 13 October 2003), the instrument in question could not be registered until it had been stamped. And if it was out of time for stamping (28 days) then a penalty would be exacted. Again, if one wanted to rely on a stampable instrument being evidence in court, it could not be admitted to evidence unless the appropriate duty (and by implication any relevant penalty) had been paid.

A succession of cases gnawed away at the underlying principles of stamp duty, pointing to a charge of duty where avoidance of payment was seen to be contrary to the substance of the transaction. In more recent years, statutory measures followed in support, such as duty being payable on an agreement for lease, rather than the lease itself, with the lease in due course being denoted with the duty paid on the enabling agreement. By 21 December 2003, we had finally arrived at a transaction tax as such, and the best way to try to understand it is, first, to forget everything one thought one knew about stamp duty.

On development agreements, one once had to take care to ensure that one did not unwittingly structure the transaction so that more duty was payable than should be. Thus, for example, a landowner might confer a development agreement including an agreement for lease for monetary consideration. His motives for prescribing development obligations might be far removed from needing the resultant development for his own use. But if there were development obligations, if the conveyancer drafted the lease as being in consideration not only of the premium (lump sum) paid but also of the observance and performance of the development obligations, the value of the built outcome would be thereby added, and the stamp duty would rocket.

The following insight may seem a little complex, but it is an example of the new world order applying to land transactions. Under a development agreement the value of the works may now be taken into consideration, but only in so far as they enhance the value of the land. So, if the landowner wants work done which does not enhance the value of the land, then it appears that it is not an SDLT matter. The value of the works is again only chargeable consideration insofar as the works do not meet certain conditions (Paragraph 10, Schedule 4, Finance Act 2003). These are:

1   The works are carried out after the effective date (usually the date of the transaction).
2   The works will be carried out on the land which will be acquired (or on other land held by the buyer or a connected person).
3   The works will not be carried out by the seller (or a connected person).

If all these conditions are met, the value of the works does not count as chargeable consideration. Thus, for example, if a development agreement is issued at a premium and then all the conditions are satisfied, there should be no SDLT on account of those conditions, and only the premium is chargeable.

However, if the seller (landowner) has an interest in the built outcome which relies on unascertained consideration (such as overage or a share in rents), there is a further divergence from stamp duty. Older stamp duty rules had strict provisions regarding unascertained consideration, and facilitated avoidance. Under the new regime, if, for example, the landowner is entitled to overage or a percentage of rents, then Sections 51 and 80 of the Finance Act 2003 make provision as to uncertain or unascertained consideration. In other words, it is a land transaction, and if there is an element of it which can be taxed, then so it will be taxed. The implication is the possibility of having to include a provisional calculation and then, perhaps, the need to file a further return to reclaim SDLT as appropriate. Here is another twist, therefore. Again taking the analogy of a house buyer, you have to fill out a tax return: your solicitor does it for you because much of the information is code based and largely meaningless unless read with the guidance notes to hand, but only you can sign it. (From December 2004, there are limited modifications allowing an agent to certify as to the effective date, and his signature suffices to obtain an SDLT certificate.) In the case of a development agreement, where contingent or uncertain SDLT will or may not be paid for at least six months after the effective date (accordingly, almost certainly in the case of development), then Section 90 of the Finance Act 2003 contains provisions for deferment. In this regard there are rules in Part 4 of the Stamp Duty Land Tax (Administration) Regulations 2003. There are provisions prescribing for providing information, including tax avoidance. However, if the application is accepted, the additional SDLT becomes due 30 days after the relevant consideration is ascertained or ceases to be contingent.

## The institutional forward funding agreement

An investor, or fund, may look beyond simply buying investment property to participation in the development itself. Indeed, many investors do precisely that in their own right and directly engage with the construction process. Developers also find the harnessing of investment funds at an early stage an attractive form of funding. The carrying cost, in broad terms, is thus borne by the fund. The essence of forward funding agreements, therefore, is that the fund buys the site at the outset and contracts the developer to carry out the development, usually paying the developer as the works proceed. Funds find this attractive because the developer will have packaged the proposal in such a way as to have investment potential from the outset and the fund engages the developer to put it all into practice.

Funds are inherently cautious, although one would not think so judging by the 1980s boom. Forward funding agreements were frequently entered into without pre-lets being in place and, by a variety of mechanisms as we shall see below, the cost of supporting the fund's outlay devolved upon the developer. Following the 1990s' recession, pre-lets were the norm albeit not exclusively, on the basis that expenditure is justified by value, but value can only be realised by a return on one's investment, i.e. rental income.

The engagement by the fund of the developer under a forward funding agreement is in some ways analogous to a design and build construction contract, and the developer is thus treated as the contractor (and indeed is also so treated for certain tax purposes). The resultant development funding agreement has, superficially, many of the same characteristics as the classic landowner development agreement. The fund's interest in the built outcome is total, however, in that it owns the land from the outset and retains and enjoys the entirety of the resultant development. All the development obligations and their accompanying performance criteria are drawn from that perspective. The benefit of the agreement will not usually be assignable in the hands of the developer: the fund is not creating a bankable instrument because it is itself paying the developer to provide the development. If the developer fails, the land and everything built to date reposes with the fund and so the developer can be dismissed. It is to be expected, therefore, that the various construction warranties required to be given to the fund will contain step-in rights. (There are a number of hybrids around. One such leaves the developer to interim fund – for example, by way of bank loan – where the development is pre-let, and the developer banks the benefit of the forward funding agreement, i.e. the payment he may expect to receive from the fund once the development is complete and income producing. Consequences of developer default thus introduce similar issues on alienation as arise in landowner/developer agreements. After all, the fund is the landowner, albeit under forward funding it remains so.)

The developer thus occupies the site solely as licensee of the fund, the developer having no interest in land as such. If the developer is in such serious default that the forward funding agreement is terminated, so also is his right to be on the land, and all of his construction contracts are similarly imperilled. The default provisions of a forward funding agreement will, as with other kinds of development agreements, include insolvency events. Default is a problematical area for lawyers to negotiate. Many forward funding agreements are extremely stringent in their terms so far as concerns events of default.

The developer's remuneration is formula-based: however, if he has brought to the table a scheme with planning permission and a pre-let, the more he has done to achieve the goal of a completed and let development, the more unfair it is if he should be deprived of his remuneration in due course. Formula-based remuneration is usually related in some way to the value of the resultant investment as against the fund's expenditure on it.

Two main documents drive the process of forward funding, first, the contract for sale and purchase between developer and fund, assuming always the developer has procured the land and is now selling it to the fund with the view to entering into a forward funding agreement. The second document, therefore, is the forward funding agreement itself. Assuming a conventional form, funding is provided under controlled conditions subject to a maximum commitment which, combined with the benefit to the fund of the completed letting(s) produces formula-based remuneration for the developer.

Insofar as the development is not pre-let, it is said to be speculative. The formula for the developer's remuneration may be as simple or as complicated as the parties may agree. For example, a single-tenant development may give rise to a formula on the basis of A minus B equals C where A is the capital value upon completion of the letting, B comprises the development costs (including a rolled up notional accumulation – see below) and C is the balance over, the whole or part of which is paid to the developer.

The developer will carry out the development under strictly controlled conditions and a glance at the development provisions of a forward funding agreement will show that the ingredients are very similar to those to be found in the landowner/developer development agreement. The perspective, however, is different because all of the performance criteria are directed towards the fund's investment at large, whereas under a landowner/developer agreement the perspective is that of the landowner's interest in built outcome, itself tempered by the fact that the developer will be taking a grant of a major interest. With forward funding, there is no such grant, the entirety of the interest already and progressively comprises the fund's investment.

Moreover, under forward funding the satisfaction of performance criteria also triggers draw-downs of funding. Commonly, these will be not less than monthly, with the developer indenting for whatever development costs he has recently incurred. Development costs, therefore, will be strictly defined and any expenditure outside of the definition will have to be met by the developer. The role of the fund's surveyor or other representative, therefore, is rather more than one of mere supervision and approval. He also has the task of approving draw-downs. Further, depending upon the terms of the forward funding agreement, the fund may reserve the right to pay items of development cost direct to, say, the contractor or anyone or more of the professionals. For a developer who is himself not substantially resourced, it may be difficult to argue with a fund about the ground rules for payment. However, from the developer's perspective, control over expenditure is desirable, if only to combine the effect of sticks and carrots in his relationships with those engaged in construction of the works.

Whilst the developer has been engaged to carry out the development, the analogy of a development agreement to a construction contract is, however, tenuous. The fund's representative, whoever he may be, is simply engaged by the fund to operate the agreement, to exercise judgement and in particular to operate the development account upon which the whole of the financial formula of the agreement depends. The developer's relationships with his contractors and professional team are separate and distinct. Thus, the funding agreement should aspire to compatibility with construction industry practice, whilst construction agreements of their various kinds should deliver institutional expectations. This of course lies at the heart of procurement and powers the merry-go-round of negotiation. (The provisions of the funding agreement, like a landowner/developer agreement, may yet find their way into the appendices of the construction contract.)

Advances of development finance would, until the development was complete and income-producing, be stale money. The fund's outlay must, therefore, be seen to be working from the moment it leaves the bank account where, otherwise, it would be earning interest. Accordingly, a notional accumulation of interest, sometimes called a finance charge, will be added to the account, periodically, thus accelerating progress towards reaching the maximum commitment. The developer is contractually not entitled to draw down beyond the maximum commitment and must, if necessary, resource the rest of the development from his own money. If he fails to do so, so also will there be a failure of performance with resultant default.

Insofar as the development is not pre-let, the fund may require an income commitment from the developer from completion until letting. The collapse of some projects in the 1990s' recession was born of the assumption that tenants would materialise. A scheme would be less imperilled if the developer (or its parent company as guarantor) was suitably resourced. Under some kinds of forward funding, known as 'profit erosion', the ongoing income commitment is reflected as a further roll-up against the developer's ultimate remuneration. In some cases there is expressly no recourse, and deals of that kind which failed in the recession rebounded entirely on the investing institutions concerned, leaving the developer to walk away with no financial commitment. Profit erosion forms of forward funding have been described as 'the perfect off-balance-sheet funding medium'. (Other off-balance sheet funding techniques, as applied by occupiers, involve a single-purpose vehicle (SPV) framed as a grantor of an institutional form of occupation lease. This is backed by long-term funding via a bank loan supported by the income stream from the lease, coupled with a tenant's option to acquire the reversionary interest upon expiry of the term.)

More than any other kind of development agreement, forward funding reflects the fund's interest in the built outcome to the extent that contractor and professional team warranties may be expected to contain step-in rights. The fund is not bound to re-employ, or continue the engagement of, the construction team but that facility subsists. All development agreements should, of course, contemplate the possibility of members of the team being replaced, because of their own defaults. As between developer and fund, however, default is extremely problematical. Most development agreements of this kind will be drawn so as to be determined upon developer default (including of course an event of insolvency). This may be harsh at any time for members of the construction team whose contracts by reason of the developer's licence to be on site at all, are effectively frustrated. Not that the developer can himself plead the doctrine of frustration by way of relief as against his construction team as a result of his own default under the funding agreement. The construction team can only hope that they will be re-engaged to finish the job.

For the developer, however, the consequences of default may be seen as extremely unfair. He will lose the prospects of remuneration, however far advanced the works. Remembering that, insofar as there is not already a pre-let, the forward funding agreement will require the developer to seek lettings, the greater the success and the further advanced the works, the more unfair it becomes for the developer. There is no perfect answer and the developer may not succeed in negotiating anything better. However, the best compromise may be that once the fund has met the additional cost of itself completing the development, and has recouped any other losses, a residual payment may be allowed to the developer.

Finally, there is the matter of disputes. As with other development agreements, forward funding agreements comprise construction contracts in lower case. Disputes cause delay, and delay also means delay to the construction contracts for the works. They are best avoided and, however unpalatable, any member of the construction team who fails to perform or who conducts a confrontational relationship with his immediate employer, may himself be the one who brings down the house of cards.

At this point, it is helpful to reflect, because the subject has not hitherto been broached in the context of development agreements at large. Underlying every development agreement is the need for insurance and for every interested party to be insured in some way against the perils inherent in any project. All properly drawn development agreements, including those mentioned in this chapter and in Chapter 4, may be expected to contain insurance provisions and these are discussed in more detail in Chapter 9.

## SDLT and forward funding

The advent of SDLT presages possible changes in the way that forward funding transactions are structured. The underlying basis of SDLT is that there has to be a land transaction. A forward funding agreement is, in essence, a construction contract (but is not treated as construction contract for Construction Act purposes, in terms of the scheme for construction contracts, unless it is wrapped up in the sale of the freehold or of a lease of not less than 12 months, as prescribed by the 1998 exclusion order).

Until SDLT, a developer could sell land to a fund for so much, and then proceed under the forward funding agreement. Stamp duty was paid on the transfer of the land, and the developer's remuneration was not only unascertained, but it fell due under the commercial contract. However, under SDLT, Paragraph 4(3), Schedule 4 of the Finance Act 2003 provides the key. If both elements, acquisition and construction, are 'in substance' one bargain, SDLT may yet fall on the forward funding agreement. Whilst guidance notes will no doubt emerge from time to time from the Inland Revenue, the new regime will test the inventiveness of lawyers as they seek to distance the two elements from each other. Current guidance notes suggest that the Inland Revenue will treat forward funding agreements much as they did for stamp duty but, in the nature of the tax, perhaps too much comfort should not be assumed. Far be it from this author to attempt SDLT avoidance schemes, but the battle lines are drawn, and the future will be interesting.

# 4 Forward sale, loan and occupational tenant agreements

## Or: And here's how the others do it

The three further development agreements described in this chapter complete the clutch of principal development agreements into which a developer may usually enter and which, in turn, may shape the developer's ultimate relationship with his construction team. It follows that only if the developer is wholly resourced, and can afford to speculate on the financial success of the development, can his contractual relations with the contractor and professional team be free of outside influence. Even then, a prudent developer must anticipate the possibility of introduction of outside contractual influences, for example the introduction of institutional interest or the securing of a prospective tenant where once the development was speculative. Within, therefore, lie further elements of the rationale for various employers' requirements in construction contracts, which, for the most part, it will be largely fruitless for the contractor or professional concerned to resist.

### Forward sale agreements

Forward funding is sometimes labelled forward sale but, for the purposes of this book, two very different kinds of agreement are contemplated. Forward funding, as we have seen, suggests the introduction of institutional funding at the outset secured by immediate acquisition of the land from or through the developer, and appointment by the fund of the developer to carry out the development and, so far as the tenants have not been contracted, then to secure lettings. Forward sale, by contrast, is essentially a conveyancing contract conditional, at the very least, on the development being carried out.

It follows that, with forward sale, the land remains with the developer during the development period. He is the legal owner (or he may of course have the benefit of a landowner/developer agreement of the kind first described in Chapter 3).

The shape and form of the development obligations will be influenced to some extent by the stage which the development has reached by the time the fund becomes involved. For example, if all statutory and other consents have already been obtained, if the building contract and professional appointments are already in place, indeed if pre-let agreements have also already been signed up, the inference is that the fund will have examined all of these and decided that they are acceptable. The resultant development obligations will thus emphasise performance criteria over initial approvals, providing in the case of the latter more for fallback positions, criteria for replacement of members of the team and so on. At first sight, therefore, the development obligations in a forward sale agreement may look very similar indeed to those in a forward funding agreement. Indeed, they may be almost identical save that the supervision process is solely in support of performance criteria, including timing, and not linked to interim draw-downs of development finance.

It further follows that the developer must source his own funding and, because he is contractually committed to sell the completed development to the fund, he must use either his own resources or loan finance.

Institutional investors are always concerned with timing, whether or not financial resources are immediately deployed. If not immediately deployed then they are at least earmarked for spending within a certain time, to make the money work for its living. The default provisions in a forward sale agreement will look fairly different from those in a forward funding agreement. If the developer is in breach, the fund can simply walk away. However, given that the principal interest of the fund is in a prospectively completed (and no doubt let) development there can be different views to be taken on, for example, insolvency as an event of default. Developer insolvency, if incorporated in an agreement for lease as an event of default, for example, may be seen as an unbankable proposition if there is a pre-let followed by a banking transaction which relies on that letting in order to realise sufficient value to repay the loan. Similarly, a bank coming into a situation where a forward sale agreement already exists, and under which developer insolvency is an event of default, must evaluate the development in terms of the prospects of the end purchaser being lost and the prospective market as a fallback. However, if a fund comes into the situation where there is a fully banked and pre-let development, and depending on how far advanced the development is, this presents food for thought for the fund as to what the consequences of default should be. The developer must negotiate what he can and hope that his advisers may influence the institutional view. The starting point is, always, that developer insolvency equals default (developer/tenant agreements apart).

With lending and institutional funders aboard, possibly an underlying landowner and any number of occupational tenants, it is hardly surprising that the related construction contracts of their various kinds must be constructed to meet a variety of needs. Given that developer default, of whatever kind, essentially means that the fund can walk away, there is much else which may yet remain intact. Say, for example, there has been a delay, or the developer has not delivered sufficient lettings under the terms of the forward sale agreement and, one way or another, the fund is not obliged to complete the purchase and has decided to walk away. The developer still retains the site and (unless the interest is no more than a development agreement or a lease which could be determined under the terms of a development agreement), it is not as if some licence to be on site is terminated. He is a landowner in his own right. Superficially, therefore, the developer is still able to proceed with his commitments to the construction team and, of course, he is legally committed to do so. Further, although the matter will be discussed in greater detail in Chapter 7, it is likely that construction warranties will not yet have been given to the fund on the basis that a condition of the forward sale will be that such warranties should be procured upon completion of the development and in any event not later than completion of the sale of that completed development to the fund. Insofar as they have already been given, and the forward sale agreement is at an end, so also is the fund's interest and so any existing warranties will be rendered ineffective.

However, unless the developer has funded from own resources, the inference is that he will have sought interim finance from a lender and the implications of loan agreements should be considered.

SDLT: given the SDLT impact on the landlord/developer development agreement, the SDLT consequence for forward sale is inevitable. It is caught fair and square. The buyer is buying a completed building which is reflected in the consideration he pays. Second, no less contrary to the old stamp duty regime earlier discussed, the rules for contingent and uncertain consideration apply, for example, in the case of overage payable to the seller (developer) on completed lettings in due course. The works will have been completed, and convincing the Inland Revenue that the further consideration may not be paid for at least six months after the effective date (Section 90 Finance Act 2003) and the Administration Regulations also earlier mentioned, may be hard to overcome. Thus, the starting point for the buyer (fund) is that SDLT will probably be payable.

## Loan agreements

In one sense, contractors and professionals should be pleased that the developer is funded and that he can thus meet his contractual commitments. From the developer's perspective, however, he has only to introduce a funder and he is no longer his own master. Contractors and professionals should always be concerned as to the resources available to the developer. A single-purpose subsidiary vehicle may be a nominal entity, and no recourse to a parent company can be assumed unless encapsulated in a guarantee, indemnity or some other kind of security. If bank finance is used, then as has already been seen, a single-purpose vehicle will likely be employed to enable the lender to exercise maximum control including the ability to appoint an administrative receiver (pursuant to a debenture) thus placing other kinds of security, perhaps to mezzanine financiers, in a truly subsidiary role.

A construction employer, perhaps building for its own use, may well have banking facilities available for its business at large. It may already have given a debenture to its usual clearing bank to secure advances generally, protect overdrafts and so on. As will be recalled a debenture broadly creates a blanket security which, in case of default, 'crystallises' and thus fixes the security over the specific assets and classes of assets generally which the debenture secures. So long as the security is effectively over the whole or substantially the whole of the assets of the company, an administrative receiver (and now an administrator) may be appointed who can, amongst other things, and apart from having the ability to operate the business of the borrower, also ward off an application by a creditor for appointment of a (pre-Enterprise Act) administrator. Likely as not, with such general arrangements in place, the most that a bank lender will take by way of security, in addition, is a fixed charge over a particular land asset, the subject of the prospective development, and operate the loan facility under its general arrangements with its borrower.

In the case of development finance, however, where the business of the borrower is essentially development *per se*, a number of security arrangements can be expected, particularly as the 'developer' will likely be a single-purpose vehicle. Certain of these directly impact on the relationship between developer and the component parts of the construction team. Again, it should be remembered that banks are in the business of lending, not development. Moreover, unless the development site stands alone, and the developer is not beholden to some third party under a development agreement which includes insolvency as an event of default, then although the exercise of the right of sale may be the only practical way to obtain repayment of the loan, a stand-alone asset also affords the bank the opportunity of appointing a receiver/administrator in case of default. Under this section of this chapter, therefore, both kinds of security and related development obligations are considered.

A loan agreement will usually be written separately from the kinds of security which underpin it. Even if the developer is resourced at the outset, it is as well for him to anticipate bank requirements should his business require it at a later stage. The provisions of any loan agreement will usually require a raft of security to be given, and it will probably look something like this. First, whatever the nomenclature, a first legal charge will be given over the land. In simple terms, this is a traditional mortgage comprising primary security, and if given by a company will require to be registered at the Companies Registry and, once so registered, also at the Land Registry. There are fixed timescales for registration in each case, and failure to register renders the security ineffective as opposed to the contractual obligations which the related loan agreement contains.

The mortgage deed will contain numerous provisions seemingly at odds with the requirements of the development process, and with the need to provide occupiers to create value in the resultant development. Thus there will, for example, be provisions about the repair of buildings, a prohibition on alterations and carrying out of works, and a prohibition on dealings, sales, lettings and so on. There will be extensive provisions concerning default and the remedies of the lender, including that the statutory power of sale becomes immediately available and a series of events of default broadly based on breach of the detailed provisions of the mortgage deed and/or the insolvency of the borrower. It will go on to provide that not only may a receiver be appointed (as prescribed the Law of Property Act 1925) but also that the receiver will be deemed the legal attorney of the borrower so that he can deal in the name of the borrower with the property as he thinks fit. Bank security documents of all kinds tend to be far-reaching and quite uncompromising in their terms. As lawyers are aware, it is usually fruitless to seek to amend these as banks will not compromise their security. The document to amend, if at all, is the loan agreement whose provisions will be expressed to prevail.

The next document to look for is the debenture, as mentioned above, which comprises a fixed and floating charge on the wider assets of the borrower, both present and future. If a debenture has already been given in the ordinary course of business of the borrower, then so soon as the borrower acquires, say, the development land, the debenture potentially covers that land also. The security 'floats' until it crystallises, as mentioned above. The debenture may be expected to contain far-reaching regulation over fixed assets and over other assets brought within the security upon crystallisation. It is of the essence of a floating charge, however, that the borrower can deal with the assets until the security crystallises. The debenture will then go on to deal specifically with default and receivership, in the case of the latter particularly as to appointment of an administrative receiver in case of default, and in case of new debentures administration under the Enterprise Act 2002. Draftsmen of new debentures must ensure compliance with Section 250 of the 2002 Act. A new Schedule B1 to the Insolvency Act 1986 is prescribed and, in particular, Paragraph 14(2) provides that new debentures must ensure that they contain a number of elements, including that the paragraph in question applies to the floating charge, that the charge purports to empower the holder to appoint an administrator, and to make such an appointment which would otherwise have been an appointment of an administrative receiver.

Land-based security can be created, by reason of statute, by a mortgage/charge as a legal entity in its own right. A throwback to the old common law exists in relation to other assets. Thus, for example, a building contract, a construction appointment and so on must all be 'assigned' under instruments sometimes labelled 'security assignments'. Such an assignment (or transfer) is precisely that, and it is only clothed with the notion of security by containing an express provision for redemption of the security and re-assignment to the original owner. A security assignment is thus distinguished from an assignment 'TO HOLD unto the assignee absolutely'. The benefit of the contract in question having been so assigned, the assignee (i.e. lender) can thus enforce it although, in practice, from the developer's perspective, security assignments are best drawn so as to contain provisions nonetheless permitting the developer to enforce the contract in question in his own right until notified of an event of default under his arrangements with the lender. Otherwise it will, one hopes, be seen as obvious that the security so given would make it impossible for the developer to carry out the development.

Security will often be taken over a number of other assets and from third parties. In the case of a single-purpose vehicle, having already bought the land with the lender's money, and having no other assets of its own, the implication for the borrower is that guarantees may be required. These may be sought not only from the parent company but in addition, or perhaps in the alternative, from one or more joint venture partners involved in the project. Thus, for example, the contractor if enjoying a profit-share agreement with the developer may offer a guarantee of the developer's obligations under the loan of transaction whilst at the same time being separately contracted as contractor for the project. Last but not least (but these will be discussed later in Chapter 7), there is the matter of a raft of warranties which, it may be expected, will allow for step-in/novation, including to the new developer in due course. Warranties in this form obviate the need for assignment, because directing or affording step-in/novation achieves much the same result, but out-and-out assignment is commonly practised. This is why many forms of appointment are drafted, away from the standard forms produced by the professional bodies to provide for assignment, perhaps with consent, but in the case of assignment to a mortgagee, without consent at all.

Much of the above does not seem to square, at least apparently, with the requirements of the development which the lender's funds are intended to support. The security, of course, is intended to remain in place until such time as the loan is repaid. Even if the facility is short to medium term, the underlying security documentation must be drawn in such a way as not to be diluted until such time as there is a disposal and the lender is repaid. The borrower, of course, has a 'right of redemption' on repaying the loan, but expectations usually are that repayment will be out of the proceeds of sale once the development is completed and occupied. Anything in advance of that suggests that there has probably been default, and that proceeds may possibly not suffice. That being so, exercise of the power of sale by the lender will not of itself release the personal debt.

The loan agreement, supported by the underlying security, is what drives the principal relationship between developer and banker. The underlying security is thus essentially the banker's fallback. The banker certainly has an interest in the built outcome in order to achieve full repayment and, to that end, to maximise the value. Accordingly, the loan agreement is likely to contain a raft of development obligations and performance criteria very similar in form, if not in substance, to those in other forms of development agreement. The rationale for those obligations is to safeguard the advances made and to secure repayment of the loan. Accordingly, by contrast with the security documents, the loan agreement prescribes for positive opposites, instead of no alterations, development, and instead of dealings, lettings, for example, and goes on to prescribe that the loan agreement prevails over the security taken.

So, one can expect to find the full raft of development obligations from consents to approval of construction team and of their terms of engagement, development and performance criteria and also supervision. Much as with a forward funding agreement, a bank may be expected to appoint a representative to approve draw-downs as the development progresses. Against the advances so made there will, of course, be a charge to interest, real interest this time (as opposed to notional interest under forward funding). If not paid directly by the developer then it will in any case be debited to the loan account together with any other charges of the lender.

Whatever the source of funding, the lender's own costs will also form part of the development costs or advance as the case may be. The lender's representative's fees are a case in point. A question sometimes arises whether, in case of forward funding or lending, other professional fees should be incurred, particularly those which track the functions of the developer's own professionals. In general terms, the layering on of funders' own professionals needs to be justified and a well-advised developer should resist unless a clear rationale can be shown. After all, members of the developer's professional team will in any case be asked to owe an express duty of care to the funder or lender, whoever it is, and to give appropriate warranties accordingly.

Because a developer has the legal right to repay his loan, a loan agreement may be expected to provide that in case of any repayment then the like amount cannot be drawn down again. Further, once advances reach the maximum commitment allowed by the lender, unless the developer is otherwise resourced to complete the development the chances are that performance criteria, including timing, will not be met and that an event of default will arise. The prior due diligence carried out by the lender will, hopefully, have considered all these elements and, particularly given that unless the developer is itself resourced, top-up or 'mezzanine' facilities may thus also be required. These are based upon secondary security (which usually requires the consent of the primary lender) and a related loan agreement. Development funds may come from a variety of sources, and every one of those sources presents a potential requirement for security including direct obligation in the form of warranties from the construction team.

Events of default under loan agreements may be expected to be comprehensive, including default under related development agreements and other related contracts. The constitution and strength of the developer itself is an important factor, whether for example its own financial strength or the skills of key people within it. Indeed, departure of a key man may itself comprise an express event of default. Once default is triggered then exercise of the lender's formal powers comes into play. It is far from entirely untrue to say that such is the stringency of formal loan agreements that, upon analysis, the borrower is probably technically in default all the time. The lender's remedies hang like the Sword of Damocles, ready to intervene as soon as the project is in the slightest imperilled. There is practically nothing that a contractor or a construction professional can do, when presented with a funding agreement of whatever kind, except comply with it so far as affecting him contractually. While loan agreements are usually the more negotiable than underlying security documents, banks tend towards subjective criteria far more readily, which is the price the developer must pay for perhaps relatively favourable financial terms. Whilst cast as an employer, he may be no more than a poodle. The more stringent and subjective the provisions the greater the fallback for the lender and the greater the risk to the developer. As has already been seen, subjectivity in other development agreements simply creates risk for lenders and so prospectively renders the project unbankable.

## Developer/tenant agreements

In all development other than for own occupation, and except in cases of outright disposal, agreements for lease with occupational tenants are to be expected. They are conveyancing contracts conditional upon development, containing development provisions the performance of which will lead in due course to the occupational lease being granted.

In major development, such agreements, particularly with the lead tenant or tenants, comprise substantial development agreements in their own right. Sometimes, they comprise the benchmark for the whole project and so the bankability of the project depends as much on the occupational agreement for lease as, say, on the prospects of institutional purchase in due course, indeed if not more so. As any banker will tell you, capital value lies in the receipt of income, and not in the bricks and mortar. If the development is for occupation by a single tenant, this will be immediately apparent. It is the catalyst for creation of the underlying investment. Moreover, unless the development, or the relevant part of it, is speculative, the prospects are that tenants will have their own particular requirements which will also tend to find their way into the building contract preliminaries and become the subject of the construction process at large.

Indeed, some developments are tenant-driven from the outset. It is often a matter of finding investors, or routing the proposal through a professional sponsor, to produce a structure which ultimately leads to the tenant and his lease being the subject of an institutional investment. Indeed, from a tenant's perspective, rules of corporation tax have not yet entirely closed the loophole on 'reverse premiums' (inducements), and if the transaction can be conducted within, say, one of the few remaining enterprise zones, or in a grant-assisted area, there is room for the entrepreneurially minded to strike an even more favourable deal.

The trick lies in the source and direction of the inducement payment. Let us say, for example, that the prospective tenant has identified a site and has costed a building for which, when let, and in round figures, an investor might be prepared to pay £10,000,000. The initial annual rent would be, say, £800,000. The land price is, say, £1,000,000 and the building costs £4,000,000, all expenses up, as the idea unfolds. A developer is brought in at a fee of say £500,000 to provide a pivot (perhaps stooge might be a better term!). The developer buys the land and funds the development with an institution. He can do this because he has a pre-let , and the whole financial structure is dealt with at once at the outset. The fund pays the £10,000,000 to the developer under a development agreement under which the developer also undertakes to procure a letting to the already willing tenant. Under a separate development agreement between the developer and the tenant, under which the fund (and the fund alone) also undertakes directly to the tenant to grant the lease (and does so), the tenant further receives development obligations from the developer (but not from the fund). The lease is granted and the tenant is on rent immediately. The fund, however, retains control of the building cost element of the £10,000,000 by placing it on secure deposit at the developer's 'request', and signs off draw-downs to the developer. Because the tenant is making payment of rent from day one, under a priorities agreement between fund, developer and tenant if the developer does not perform then the fund may do so, and if the fund does not do so then the tenant may do so, accessing the reserved funds for the purpose.

Meanwhile, the surplus, over and above land cost, building cost and developer's fee, is paid by the developer (not by the fund because it has already been paid to the developer) to the tenant as the inducement, under a further separate agreement. That sum should come free of corporation tax. Why? because no capital sum has been paid by the landlord (fund) to the tenant. Had it been so, it would have attracted corporation tax. Because of the tenant's unique position, access to institutional funding is the key (and a single-tenant development so structured is one of the rare occasions when warranties in favour of the tenant contain step-in rights).

However, developer/tenant agreements are perhaps usually less sophisticated and concentrate the more on tenant requirements as such. Unless there has been a pre-let, a developer must anticipate tenants' requirements and negotiate all his related development agreements, including funding transactions, not to speak of construction agreements, in line. Nevertheless, the negotiating power of the tenant is considerable in a tenants' market and that power can place considerable strain on pre-existing relationships.

Where a fund is involved at the outset, a developer should ensure that the fund enters into the agreement for lease at the very least to grant the lease, without which the best-laid strategy may yet fail. An error sometimes made by developers, or rather by their lawyers, particularly where the fund is not directly involved at the outset, is also to fail notwithstanding to make development obligations personal to the developer. An agreement between developer and tenant will likely, despite denials therein to the contrary, have legal effect to combine in the relationship of landlord and tenant the original developer's obligations, unless those obligations are made expressly personal to the developer alone. That those obligations should be visited upon a purchasing fund may endanger the transaction at large.

Much as with forward sale and other development contractual media, a tenant is concerned not just with performance but also with timing. Whatever other remedy he may have, failure to deliver on time at least provides him with the ability to walk away. The hapless construction professional may have a valid defence for his own breach of warranty but attempts at ring fencing his exposure will likely be resisted.

Developer/tenant agreements lie at the heart of investment in new buildings. Whilst not perhaps of immediate concern for those involved in construction, it should nonetheless be noted that because of this pivotal function, all principal forms of development agreements, particularly those discussed in this chapter and the preceding chapter rely on what is known broadly as 'letting policy' in the absence of specific pre-lets. The criteria for letting go to the heart of the investment or, in the case of bank lending, to maximisation of value with the view to ensuring payment of the loan. The letting criteria will be whatever is perceived as appropriate to the circumstances. However, the more lettable units there will be, say in a shopping centre, the more difficult it becomes to devise a policy. An example comes to mind, pre-1990s' recession, of a shopping development where joint agents were appointed by the developer, one of these at the instance of a fund pursuant to a forward funding agreement, being also the fund's surveyors in that capacity. For the purpose of the exercise, nevertheless, both firms were the formal appointees of the developer. Between them they devised a letting policy for the shopping centre and between them they both understood, or thought they understood, and indeed individually they did understand, precisely what that letting policy was. In the event, when prospective tenants came forward they could not agree on how the criteria they had prescribed actually applied in context. The result was a slow take-up, and it does not take too much imagination to appreciate that delay adversely affected those tenants who had already arrived. The last one to turn out the light was the local authority's tourist and information centre which had taken up a unit at the entrance of the complex.

The happier side to the story is that everyone involved in the construction was duly paid, the transaction being structured around speculative (spec) forward funding, but there was also a lesson for this particular author as developer's solicitor. Having been 'ganged up' upon by his own client, the fund and the fund's surveyors, so that the lawyers on both sides were forbidden to interfere, one was not able to test the practicality of what was being said. Never will that error be made again, however unpopular one may become. Every part of a development agreement has the potential for being minutely examined by counsel and judge, and every nuance dissected. After all, that is the way they earn their crust. The same thought should perhaps thus also surround building contract provisions, about which more later. If a proposition cannot withstand a single interpretation, then dispute beckons, and when that happens, only the lawyers can win.

## SDLT

Developer/tenant agreements, of course, contain construction obligations. Again, anything paid by the tenant in relation to those obligations will be within the charge to SDLT if both elements, that is to say lease and construction, are in substance one bargain. However, what is paid will, in most circumstances, be the rent itself, and if that is what is paid, then SDLT follows accordingly. Again, guidance from the Inland Revenue may be expected from time to time. Incidentally, it is beyond the scope of this book to go into detail such as the calculation of SDLT on commercial leases. The formulae laid down do not make comfortable bed-fellows for the mathematically less gifted, but the Inland Revenue has a website which incorporates a calculator to assist tenants' solicitors in calculating the charge.

In those cases where the consideration is uncertain or contingent (unlikely in the case of building works preceding a full repairing and insuring (FRI) lease), then possibly deferment may be available. If the tenant is carrying out construction works for the landlord, then that is part of the consideration and the value will be taken into account.

# 5  Site assembly and elements of land law

## Or: What have we here?

'What possible business is it of mine?', an architect or engineer might ask, when first instructed to consider the scope for development and its associated infrastructure. The sentiment is directed, not to the implementation of his instructions, but to consideration of other factors which affect the ability to proceed, even though paper title to the land may be 'clean' without trace of any adverse interest, perhaps with planning permission and other statutory consents obtained, and whilst there are apparently no legal inhibitions. Unfortunately, that is far from all. A combination of the eagerness of the developer and failure of legal due diligence may yet transform the role of the construction professional to that of messenger, if not grim reaper.

There can be no substitution, of course, for the development lawyer donning wellington boots (or indulging in off-roading, if he must) to inspect the target site, and indeed surrounding land, for adverse characteristics. Better still if, for example, he is aware of existing infrastructure, particularly that relating to surrounding land. To carry out the exercise in conjunction with some suitably qualified person, such as an engineer, can be invaluable.

This author recalls inspecting a site about twenty years ago, which had ceased its usefulness as an aluminium smelting works. The orange and blue ponds and puddles were intriguing, but not as intriguing as various pumping stations, pipes (some supported above ground) and other trappings of infrastructure. These had existed for very many years and some, but not all, were reflected in a series of old unregistered conveyancing documents, to illustrate not only the installations themselves but also the water and drainage systems serving both commercial and residential developments surrounding this thirty-five acre site. A careful exercise in conjunction with an engineer, and much note taking, brought the old deeds to life, providing answers to questions which had even baffled the seller's solicitors, and much more.

Unhappily, not everything creating legal rights and obligations is to be inferred from legal documents. Legal impediments can also arise by statute, and (particularly 'prescriptive rights') by implication of law and, by no means rarely, intelligent site inspection may raise issues which, upon examination, should already have discouraged the developer from proceeding with his acquisition.

There is no substitute for competent legal due diligence, but wiser developers will be exploring with members of their intended professional team the development possibilities and the practical hurdles that may lie in the way. In the event, planning may prove to be the least of them, but this chapter is about site assembly and legal constraints affecting the land. Moreover, where legal constraints are to be found, wiser developers should encourage dialogue between solicitors and members of the professional team (and be prepared to pay for it). This chapter concentrates on some of the things which should be uppermost in the solicitor's mind as he proceeds, his developer client also.

## Site assembly techniques

An architect, for example, may be told by the developer, 'I've just bought this', or 'I've an option on that', or 'I've a first refusal' and so on. The concept of an estate in land, whether freehold or leasehold, has already been mentioned. That halfway house, 'contract', is a rather more nebulous concept. Once upon a time lawyers used to speak of a 'Section 40 memorandum', that is to say a document setting out the terms of the transaction 'signed by the party to be charged' in the parlance of the particular section of the Law of Property Act 1925 under which a contract, whilst existing in law, would not otherwise be enforceable against the person concerned. Fast forward to Section 2 of the Law of Property (Miscellaneous Provisions) Act 1989 and the validity of the contract itself required to be underpinned by the signatures of the parties to it, not just one of them. At a stroke, the need for cautionary words such as 'subject to contract' on a document, even on a letter setting out terms and signed by or on behalf of a prospective party, could largely be abandoned (although estate agents, not being lawyers, frequently use the term in correspondence the subject of which has in most cases little prospect of creating legal relations. It still has its uses in appropriate circumstances, however).

## Contracts

A contract for the sale of land creates an 'equity', that is to say a right to buy upon the terms of that contract. The contract will be imperfect if it is not entire, and modifications to it, particularly since 1989, must also be given for consideration (or it must be executed), concepts again earlier discussed, and then signed or executed by both parties. There is nothing, of course, to prevent the direct transfer/grant of land without a prior contract if the parties so wish. A contract may be subject to any (lawful) conditions that the parties agree upon. Every now and again case law intervenes to define the effect of conditionality, for example in certain instances that the parties must act in good faith, without actually saying as much. Thus, if the contract is conditional upon planning permission being granted and, say, an application was made by the seller before the contract was entered into, it is implied in law that the seller must not therefore withdraw that application. Recourse to the courts can be avoided simply by clarity and completeness, of course, and good drafting 'tells the story'.

## Options

Options have characteristics of their own, whether there is a call option (the buyer has the right to call for the property) or a put option (the seller has the right to 'put' the property onto the buyer, and so to receive payment for it accordingly). Contracts of this nature came under particular scrutiny in the early days of the 1989 Act, not least in relation to provisions for exercise, whether both parties should be required to sign any procedural documentation, whether indeed the day of the option was now over. In the event, the giving of a procedural notice of exercise was held to be merely the satisfaction of a condition, and did not go to the root of the contract itself. Depending upon the nature of the contract, whether it be deemed an option or a conditional contract in context may be academic in practice.

Exercise of an option may itself be subject to prior express conditions. For example, a buyer may take an option to buy land conditional upon his obtaining planning permission within the confines of certain parameters. He must act in good faith, but if the price to be paid to the seller also requires calculation in terms of the planning permission so obtained, the likelihood is that there will be express and stringent conditions overriding any general principle. Failure to perform may result in the option not being exercisable, and difficulties with such contracts often come down to sufficiency of the document itself.

Options have the particular characteristic that they prospectively fall foul of what is known as the rule against perpetuities, and even a contract for sale and purchase is so tainted. An option contained in a lease, in favour of the tenant to acquire a reversionary interest, say a new lease or the landlord's reversion, is in general terms prospectively immune (Section 9(1) Perpetuities and Accumulations Act 1964). An option to acquire for valuable consideration an interest in land is subject to a perpetuity period of 21 years. Put options have difficulty in being conceived as options to acquire, and so the thinking seems to be that put options, conferring no interest in land as such until exercise (at which point the buyer acquires an equity in the land) are sidelined into Section 1 of the 1964 Act which avails one of a perpetuity period of up to 80 years. One textbook writer unhelpfully observes that the sensible drafter will limit the exercise of a put option to 21 years beginning with the date of the put option agreement. We lawyers live in a quaint world of arcane theories and practices, but there are times when even the rule against perpetuities emerges to interfere with a perfectly sound development scheme.

So, when a developer prescribed that in accounting and lending terms a 30-year lease needed to contain put and call options for a new 30-year term or, in the case of the put alone, for the tenant to acquire the reversion on terms at the end of the lease term, one was presented by a dilemma. The accompanying legal paranoia produced a developer/tenant agreement for lease, incorporating development provisions, the lease itself prescribing the following:

1   Developer grants a lease of only 21 years to the tenant.
2   The lease contains put and call options respectively as regards a further nine years, thus taking the total up to 30 years in all.
3   But, within the nine-year term, first there is prescribed a call for a further 30 years. Second, also within the nine-year term is prescribed a put option, overriding the tenant if it has first exercised its call and, in any event, putting a reversionary interest onto the tenant, on terms.

Finally, options cannot pass without mention of SDLT. Options at large are separate transactions from the interests which exercise may confer. Usually, the consideration for the option will be too low for this to be of significance, but the distinction should nonetheless be noted.

A more nebulous concept still has been the right of pre-emption. In the simplest terms, it amounts to a legal right of first refusal. Under such an arrangement, encapsulated in a formal agreement, the seller says to the would-be buyer that he will not sell the land to any third party without notifying him of the terms of sale to the third party. The would-be buyer is given the opportunity to match, i.e. offer to buy upon those terms, and if he does a binding contract ensues. As before the arrangement must at the outset be committed to writing and signed by both parties.

Site assembly is a problematical area for developers and for landowners who wish to bring land forward for development. There may be tenancies to terminate, for example. Business tenants who have the protection of the Landlord and Tenant Act 1954 have certain statutory rights which can be undone upon satisfaction of a number of criteria including the intention to demolish or reconstruct. Often, it will be a matter of painstakingly buying in pieces of land, sometimes taking options where there is uncertainty, and so on. Once land is bought, title is registrable and registration is a matter of public record.

'Back-to-back' conveyancing, once beloved of property dealers, is particularly difficult now with the demise of unregistered conveyancing and the placing of registers in the public domain. Time was when, in buying a piece of unregistered land, one entered into a contract to buy the land based upon a conveyance which comprised a 'root of title' (since 1969, and the Law of Property (Miscellaneous Provisions) Act of that year of not less that 15 years standing). Post-contract, only then was full title disclosed and upon satisfactory replies to 'requisitions on title' the matter was concluded. How much easier it was, then, to buy a property whilst at the same time selling it on to someone else at a higher price. Now, in practical terms, title investigation is largely up front, the registers are open and ownership is known and, indeed, contracts for the sale of land often declare that the buyer has had full disclosure and that he is not entitled to raise requisitions save as to any matter registered after the date of the contract. (Since 13 Otober 2003, consequent upon the Land Registration Act 2002, title is not just a matter of public record, but that record alone will suffice, and a land certificate will no longer be required.)

This is not to say that contracts need not be confidential. Indeed, confidentiality clauses are common place although their value in terms of remedy for breach is debatable. Of course, some things cannot or should not be kept confidential, at least between the parties, such as accounting, taxation and stock market requirements. Moreover, whilst any contract for the sale of land creates an equity, to be binding upon a 'purchaser for value' from the seller, an option or a conditional contract must be registered against the seller's title, and confidentiality is impaired accordingly, and confidentiality is further eroded by the Land Registration Act 2002.

The registration of a 'caution' at the Land Registry, by a buyer under a contract, is (or was, see below) itself of little value as, in substance, it is no more than a flag or warning, and the buyer may be left to exercise such rights as he has against the seller. To make a contract enforceable against successors in title to the seller requires registration by 'notice', a technical term in context. The contract will contain an application to the Land Registry by both seller and buyer, so that the title becomes encumbered by the contract, subject to the priority of anything else previously registered. An unfortunate side-effect of the security thereby afforded to the buyer is that, in the nature of notice, the contract becomes a matter of public record, and so a confidentiality clause counts for nought.

The Land Registration Act 2002 came into effect on 13 October 2003. It does considerable violence to the practice of registered conveyancing developed by the Land Registration Act 1925. Gone, for example, is the caution mentioned above and new forms of notice are prescribed. There is corresponding recognition of the need, at times, for the preservation of confidentiality, and mechanisms have been devised accordingly. A significant shift of emphasis is on the positive erosion of 'overriding interests', being legal rights and interests which have otherwise escaped registration and which, to a large extent, can only be discovered on enquiry. Included in these, at least to date, have been leases for less than 21 years. Compulsory registration of leases now comes in at seven years. Moreover, even if the lease in question is not registrable, rights granted by a lease, e.g. rights of way over common parts, must be registered against the landlord's registered title in order to have legal effect. Thus, the emphasis is on a gradual change in the character of registered entries so that, in the fullness of time, everything of significance will be shown on the registers. For example, leases are now compulsorily registerable if granted for a term of as little as seven years. If a lease is granted for less than such a period, it is not registrable, but any rights associated with it, say a right of way over common parts, must be registered against the landlord's title as, otherwise, upon disposal of the landlord's reversion those rights will evaporate. Meanwhile, as regards existing leases, the new procedures prescribe for details to be lodged on disposal of the reversion. The intention is that in the fullness of time, the registers will be seen as comprehensive. The removal of one curious anomaly is welcome: non-lawyers may not realise that, hitherto, if one searched the registers of a leasehold title the one document that was not available as of right was the lease itself. We have, therefore, come a long way, particularly if one recalls that until as recently as 1990 registers were closed except to registered proprietors so that, on every conveyancing transaction concerning registered land, a written authority to inspect was required, either from the registered proprietor himself or from his solicitors on his behalf. It remains to be seen how well the Land Registry has prepared itself for the prospective avalanche of applications concerning routine matters which have hitherto not required registration. One sees difficult times ahead as Land Registry staff grapple with their creation and practitioners condition themselves to new practice and procedure. So far, it does not seem to be all bad but there are occasional instances of more complex transactions where applications can still take several months to process. This is not the norm, however.

The above apart, the established system of Land Registry searches, however, colours the picture a little more. Without disclosing his contract as such, or indeed without having one at all, a buyer can make his search which gives a priority period during which his application (assuming the seller has meanwhile executed a transfer or lease in his favour) has priority over any other application. The results of a search, made against the title at a given point in time of which the buyer will be informed, will disclose any priority interests which may in the meantime be the subject of applications. If anything new is disclosed which the seller cannot remove or will not remove, it suggests the prospect of breach of contract (depending on its terms) and the buyer declining to complete the contract.

## The lie of the land

Unfortunately, one of the commonest errors in conveyancing is that of failure to assess the extent of the site, in the sense of the buyer's expectations on the one hand and the reality of title on the other. Agents' plans should always be regarded with suspicion. In the days of unregistered conveyancing, where land was often defined without reference to a plan at all, it could change hands many times over without error, simply because the physical boundaries were well-established in relation to equally well-established surrounding ownerships. A postal address for example could work well for decades. Once well-known characteristics are removed, walls, fences, buildings and so on, the extent of ownership becomes obscure. With land registration now compulsory, bit-by-bit the uncertainty is overcome. Still, if the solicitor seeks instructions based on the filed plan to the title, only to have his client confirm acceptance, he can hardly be held negligent. The client should share it with his professional team too!

In cases of extreme difficulty in defining boundaries, other evidence may be required, including perhaps in the form of sworn declarations. Once title is registered, the 'guaranteed' nature is often misunderstood. Registration fixes you with what you have, not with the quality of what you have (save to the extent of category of title, absolute, good leasehold and possessory) with the result that analysis is no less required. Boundaries are thus but one element. Below are a few of the more common areas of difficulty.

First, say, a development site is being created within the boundaries of a much larger area of land to which, no doubt, title has already been registered. The plot is well defined in the plan produced by the seller, no doubt prepared by his surveyors. When the land is bought by the developer, the Land Registry requires that it be bought by reference to a plan clearly showing the land in question. For registration purposes, the Registry generally do not find favour with land described 'for the purposes of identification only', but the resultant filed plan is effectively no more than that unless one makes a special application. Moreover, it is by no means impossible for plotting errors to be made by their own surveyor, and when the title entries are returned to the developer's solicitor, the plan should be examined with care. Upon experience, for example, a straight line boundary on a large setting out drawing used also for conveyancing purposes was clearly shown to be curved on the 1/2500 scale Land Registry filed plan. It was not much of a curve and might have passed unnoticed by the casual observer, but actually represented an inaccuracy of several metres.

When a site is being assembled, there is plenty of scope for boundaries, particularly with unregistered titles, not fitting perfectly. Whilst prudent developers are used to having site surveys carried out, they are sometimes less used to the surveyors working closely with the solicitors to see not only that the legal title and the measured site correspond but that the component elements, upon amalgamating various legal titles actually fit. The problems of ransom strips, as well, are legion. Less often remembered is that ransom strips can be, and not infrequently are, contrived, even by public sector bodies. This author recalls precisely that occurring some 30 years ago in relation to a new town corporation. Ransom strips, wherever they are, are thus by no means the sole product of error or miscalculation. Sometimes they are deliberate and most difficult to spot. A line on a drawing, a multiplicity of colourings on plans and so on, can all contribute to the illusion of completeness, and are particularly difficult to assess. There is no general principle of law to impose upon an owner of land an obligation to dispose of everything necessary to make a site accessible, even where that owner is also the seller of a purported development site or the relevant part of it. Care in identifying the land in relation to surrounding holdings is therefore of the utmost importance. Where the existence of a ransom strip is obscure, what cannot be ascertained from plans may perhaps find remedy, albeit perhaps limited to damages, if there has been efficient due diligence through the medium of 'enquiries before contract'. The courts will nonetheless be slow to undo a transaction. Moreover, replies to most formal enquiries are usually hedged around with appropriate qualifications. Very occasionally, and such instances must be regarded as being truly rare, there will be wilful deception and the extent of available remedies must be examined in context.

Ransom strips have long been used to afford a seller a later advantage. Thus, for example, where access from the highway is essential, specific enquiry is required to establish whether a particular boundary abuts the public highway for the entirety of its length and, even though the best likely remedy may be damages, the burden of proof upon a buyer is great indeed. The maxim 'caveat emptor' is as strong as ever, a buyer is on enquiry and, worse, a subsequent finding of a ransom strip by no means points to negligence on the part of his professional advisers. Never assume that because something is wrong it is the product of negligence. Negligence always carries a heavy burden of proof.

Finally, a word of warning. Those who have studied land law will be aware of the concept of a 'way of necessity'. This will not overcome problems with ransom strips unless it is the landowner who sells the ransom strip and retains the principal site instead. In other words if you sell land that leaves you landlocked and without an exit, only then may a way of necessity arise.

## Physical characteristics

Assuming the boundaries are correctly established, and whatever legal inhibitions affect the paper title, one thing the lawyer cannot establish from title deeds is the physical characteristics of the site. Aspects of environmental law apart, enquiries about physical characteristics should be treated with care by any seller as well as by the buyer. If there are characteristics, e.g. sewers and pipes, not documented which the seller knows about, it may be or at least appear helpful to provide a helpful answer. A well-advised seller should, however, always qualify his answer. It is not enough that he should believe that his answer is correct but that it is correct. If he says he believes it is correct the implication may be that he himself was on enquiry and can justify himself. If he cannot, he must make that abundantly clear. Well-crafted replies to enquiries should in such circumstances place the burden clearly on the buyer to rely on his own inspection and enquiries of others.

Site and ground surveys are a *sine qua non* and a prudent developer should also make surveys of that nature available to the solicitor in case anything arising out of the exercise should properly signal a line of legal enquiry.

For the solicitor engaged in acquiring a development site, there is much to be learned, in the nature of physical characteristics, from routine separate enquiries of the utilities. There may, for example, already be an electricity board lease in place perhaps relating to a substation, cable rights, access and so on. There may, however, be no such documentation. Statutory aspects are dealt with in the next chapter and 'prescriptive rights' later in this chapter. What is physically in and about the site may be assessed from such enquiries, but by no means always. It is not unheard of, as one experience showed, for what was very obviously a mains cable, very much alive, simply not showing on electricity board records at all. (Best left alone by the contractor, and it duly was!)

Information available from the Environment Agency, although hedged around with qualification, is much more useful than it used to be, and is accompanied by helpfully keyed drawings. Whether the site is liable to flooding or has other relevant physical characteristics, there is no shortage of source material, still less of professional help of various kinds, to establish the suitability of land for development. But merely because, say, environmental issues are fed into the planning process provides no inherent underlying protection. *Caveat emptor* is still paramount, and one should not expect any remedy to lie against the planning authority.

Where there are mines, solicitors should make mining searches, and the proximity of railways, particularly underground railways in London, also raises issues of developability. Working in and around a railway environment above or below ground raises issues of its own, not least related to safety, particularly interference with electronic equipment and so on. There are plenty of lines of enquiry available to the buyer's solicitor in this respect. Moreover, an old mining search should not be relied on, even if mining operations ceased long before. Last week's subsidence may be hot news. The express exclusion of mining rights from legal documents requires careful examination. An exclusion of mines and minerals, for example, simply means that the buyer cannot benefit from them. It does not mean that the seller can do anything with them either. To achieve that, he must have an express right to 'win and work'.

It is also helpful to consider under this heading the physical characteristics of adjoining and surrounding land, rights of support and protection apart. The more confined a site, chances are that it will be difficult to work without infringing the rights of others. The public sector is also concerned: highway authorities will inevitably be interested in traffic movements, and safety of the public at large, erection of scaffolding outside the site, say on the pavement or out into the road, and so on. All of these matters will require consent.

If one erects a crane, swinging the boom over adjoining land is a trespass at common law. The basic characteristic of the legal title to land is that it reaches as high up into the sky as it is possible to imagine and downwards to the centre of the earth. The unlicensed over-sailing of a crane constitutes the tort of trespass which is actionable accordingly. Expect, therefore, to have to negotiate a crane over-sailing licence (the detail of which should be left to the solicitor whose enquiries may lead him to call upon not just the freeholder but tenants and others as well). The need for a crane over-sailing licence should not be under-estimated. If there is a trespass, the adjoining landowner has every right to injunct and the likelihood is that he will succeed. A brief word about injunctions then: first, they cannot be obtained outside of proceedings. Proceedings, such as in the case of crane over-sailing for trespass, are the key, and an injunction is a protective device. If obtained in an emergency, without the defendant necessarily being present, the court will extract an indemnity from the plaintiff to protect the defendant should the injunction be found not to be justified. A further hearing, at which the defendant can be heard, will determine the position until the main action is heard, and whether or not the main action upholds the injunction is another matter again.

This chapter is carefully headed with 'elements' of land law. Many more topics will be found in specialist works. The next chapter looks at a number of statutory issues, and suffice it to say here that statute has, over time, been intervening more and more in matters concerning physical characteristics, whether it be the contaminated land regime imposed by Part II of the Environmental Protection Act 1990, archaeological characteristics or indeed the statutory listing of buildings, all of which are inhibiting factors but also are reflected by information reflected in registers maintained by the local authority. Look also to planning policy guidance (PPG) notes, influencing the planning process, such as PPG16 in relation to archaeological antiquities and PPG25 relating to development and flood risk, in case of the latter particularly following the winter of 2000/1.

## Covenants

Outside of legal documents, use of terms such as 'covenants', 'restrictions', 'obligations' and so on may not seem to suggest much if any difference between them. Within legal documents, not so. Earlier in this book we considered contracts generally and distinguished the concepts of simple contract requiring legal consideration and also the doctrine of covenant. Covenant is, essentially, a one-sided obligation opposite which no sufficient corresponding obligation is required, and it also enjoys a statutory limitation period of 12 years as opposed to six years for simple contract. A covenant is, necessarily, a written document or a clause within a document written as a covenant. A document containing a covenant is a 'deed', and it is execution as a deed which identifies a covenant, not use of the word itself. Indeed 'covenant' backed by mere signature, without indication of a deed, is no covenant at all. Look for such words as 'executed as a deed' as opposed to 'signed by'.

Covenants are essentially personal obligations, that is to say enforceable only against the person who gave the covenant, i.e. the covenantor (but see below as to restrictive covenants). The person or persons expressed to benefit from the covenant (covenantee) are essentially the only ones entitled to enforce the covenant, but whether they can benefit is another matter. However, where the benefit 'touches and concerns the land' a transferee or assignee of the land can enforce it. Not to have the benefit of the subject matter by contrast implies no loss, in turn suggesting no liability. Where covenants are involved, determining precisely who is burdened by and who may benefit from them will usually require legal advice unless the intention is clear, and covenants relating to land are certainly no exception.

Enforceability is thus an immediate problem area, and the lawyer starts by identifying the original covenantor and covenantee and then determining on the wording of the covenant who may benefit from it and who is burdened by it. Thus, a covenant to maintain a fence may be expressed to benefit not only the adjoining owner in question but also his successors in title, in which case those successors will be able to enforce it, but it doesn't have to. A covenant that does not touch and concern land may not be assignable at all, unless the benefit of the subject matter expressly moves. Often it comes down to legal drafting and the related law is beyond the scope of this book.

However, a covenant essentially to do something, as opposed to a covenant not to do something, is in general terms only enforceable against the original covenantor. The person who gave the covenant to fence thus remains liable after he has sold on his land, and his own successor has no direct responsibility to perform a like obligation and neither can the adjoining owner enforce against him. Accordingly, given that the original covenantor is on the face of it alone responsible, in such circumstances he should upon disposal of the land seek an indemnity from his successor in order to provide protection for him against breach of an obligation over which, in practical terms, he has no continuing control. The burden of the covenant, is thus not generally transferable but, of course, to every rule there will often be found an exception.

First, the nature of positive and negative covenants needs to be understood. It is a matter of interpretation in every case, but no amount of negative wording can undo an obligation which is positive in substance, and vice versa. An obligation to do something 'only if ...' may thus be read in context to mean 'not to do something unless ...'. Such matters of interpretation rely on case law and legal advice may be essential.

Restrictions or restrictive covenants are loosely termed as being covenants of a negative nature. A true restrictive covenant affecting land is not only negative in nature but can also be enforced against successors in title to the covenantor so that in substance the land itself is burdened by the covenant. Although the burden of a covenant does not run against successors in common law, it can be made to do so in equity so long as certain principles are satisfied. The first is that the covenant must be negative in nature and, as has been seen, it is the substance, not the form, which counts. Second, the covenantee must own the land for the protection of which the covenant is made. Thus, if the covenantee has no land which can benefit from the protection of the covenant, then however negative in substance it is not enforceable against any person other than the person who gave the covenant i.e. the covenantor. Related to the last principle is the idea that the covenant must 'touch and concern' the land that is intended to benefit. Lawyers refer to this land as the 'dominant tenement' (and conversely the land of the covenantor as the 'servient tenement'). Whether the covenant benefits the dominant tenement is clearly a question of fact in the particular circumstances.

The fourth and final rule is that it must be the common intention of the parties that the burden of the covenant runs with the land of the covenantor;

for example, if a contrary intention be shown, the covenant cannot be a 'restrictive' covenant. However, since 1925, Section 79 of the Law of Property Act of that year prescribes a legal assumption that the covenant is to bind successors unless the contrary is expressed.

A restrictive covenant has a further statutory hurdle to cross before being established as such. Since 1925, with unregistered land the covenant needs to be registered as a Land Charge Class DII, and with registered land the covenant must be registered on the title so burdened but, conversely, the benefit is not registrable and a person may be unaware that he has the benefit. Where change of use leads to nuisance, it is a good idea to check the title of the land in question. You never know, you might have an additional line of attack.

Restrictive covenants, reaching down as they do to successors of the original covenantor, naturally burden leasehold property and other subsidiary interests as well, but what of covenants in leases as such. The essential nature of a lease is that it is one of the estates in land comprehended by statute, that is to say the 'term of years absolute'. One need not split a hair because a lease is for a term other than for a precise year or years. Rather, the distinction is between that and, say, a tenancy at will. It is beyond the scope of this book to examine issues such as whether a relationship is a lease, or not a lease, whether it exists in law or in equity and so on. Better to switch to a book solely on land law. Suffice it to concentrate here solely on a lease created by deed which contains a variety of covenants. At common law, as between original landlord and original tenant, whatever covenants they enter into, whether positive or negative in substance, may be expected to be binding accordingly. Moreover, such is the contractual nature of a lease that, prior to 1996 when the Landlord and Tenant (Covenants) Act 1995 can into force, the original parties were bound so that, however many changes of landlord and tenant, the original tenant remained personally liable even though he could not benefit from the lease. Conversely, beyond the original lessor and lessee, in the hands of successors only those covenants in the lease which touch and concern the land itself will remain binding although, in general terms, the landlord and tenant for the time being are in any event bound by its provisions. Thus, if the lease contains no provisions for fresh obligations to be given by an assignee, covenants which do not touch and concern the land will fall away. On experience, from time to time in the past, local authority landlords would attempt to impose covenants of an essentially political nature which, if there were any restrictions on alienation which would have afforded the opportunity to seek fresh covenants, would have to be negotiated out.

The 1995 Act (see above) further modified the relationship between landlord and tenant by restricting the obligations of the parties to leases entered into after the act came into force essentially to the time during which the tenant was tenant under the lease with the ability on the part of the landlord, if he so prescribed, to require the assigning tenant to maintain responsibility only during the period that his assignee was tenant and so on.

Unfortunately, there is no precise corresponding provision benefiting outgoing landlords, but there is a mechanism for release which, as structured, effectively requires express provision for release of the outgoing landlord within the terms of the lease.

Covenants of all their respective kinds require careful legal analysis as they may have effect to preclude development of the kind or in the manner which the potential of the site otherwise suggests. A seller of development land may enhance its value if he can find ways and means of limiting the impact of restrictive covenants. Whilst a restrictive covenant is in force, the burden may be reflected in the covenantee's ability to injunct. A positive covenant, in this nature, is not so tainted but may lead to an action for a specific performance. In the case of an obsolete covenant or one which would 'impede some reasonable user of the land for public or private purposes' (Section 84 Law of Property Act 1925) it may be lifted upon application to the Lands Tribunal whose powers nonetheless include the imposition of compensation as well. The applicant may find, in the alternative, that the tribunal adds further restrictive provisions instead which, if not accepted by the applicant, mean that his application will fail.

Finally, merger of titles implies extinguishment of covenants and so, in the case of development land, a site assembly exercise may have effect, in the particular circumstances, to undo a whole raft of provisions affecting different titles. However, that is only the case in relation to unregistered land because registered titles, and everything registered on them, remain separate and distinct until there is a merger of titles. It is therefore idle not to merge the titles because if, for example, an area is comprehensibly redeveloped, the new owner/occupier of one part may once again enforce covenants against the owner/occupier of another part if the covenants remain subsisting. Happily, this must be seen as a curious circumstance born of carelessness, if it should ever arise.

## Easements

Just as covenants have the potential of merger upon site assembly, so also do both express and implied rights. The law of easements is substantial and a specialist work is suggested for their study. However, legal rights can exist over land, outside of the confines of contract, whether simple or covenant, and notwithstanding that the person having the rights does not own the land or any right to own it. At an early stage in site assembly, the solicitor will review all the rights expressly benefiting and burdening the property. An easement is, in essence, a right granted out of land for the benefit of other land and, if created by deed (or by prescription – see as to the Prescription Act 1832 below), is enforceable against anyone having an estate or interest in the land, and whose conduct or omission obstructs or impedes exercise of the right. Lawyers refer to dominant and servient tenements in context, the term 'tenement' having nothing at all to do with the law of landlord and tenant.

Lesser easements are not legal easements but equitable easements, and are not enforceable against a bona fide purchaser (of the servient tenement) without notice. The lesser form is registrable as a land charge class D (III) in the case of unregistered land and if not registered either there or at the Land Registry as the case may be, it is void against a purchaser of the legal estate for money or monies worth. An equitable easement is defined as 'any easement, liberty or privilege over or affecting land and being merely an equitable interest'. Already, perhaps, the distinction may be a little too fine for a book of this nature, the important principle is that it is possible to distinguish the extent of the burden as well as the extent of the benefit.

Legal easements are, for the most part, created by deed, and they will be apparent on the face of the title deeds in the case of unregistered land and clearly shown on the title registers in the case of registered land, and unlike restrictive covenants, benefits and burdens are recorded. Rights of this nature include, for example, right of way, support and protection, light and so on. Moreover, it is in the nature of easements (including prescriptive rights mentioned below) that nothing need be done to continue their existence merely because either the dominant tenement or the servient tenement is sold. This automatic acquisition is enshrined in Section 62 of the Law of Property Act 1925 except in so far as the conveyance (or today transfer, in any event) may impose limitations. Thus, a privilege hitherto enjoyed shall be deemed to pass and may, as it were, be credited to the requirements for satisfaction of the rules for acquisition by way of prescription (again see below).

An owner cannot have an easement over his own land. He may acquire the dominant and servient tenements, but, nevertheless, keep the rights alive only by reason of failing to merge two registered titles. Once merged, always merged until a transfer of part only.

The clearest evidence of the existence of an easement is of its creation by deed, and thus also its modification or extinguishment. Sometimes, an easement may exist by deed but, if the physical characteristics of the land are such that it can no longer be exercised, then the right must die. Similarly, if a right could be exercised but is not exercised over a period, it may be deemed abandoned. There is plenty of case law on the subject and the position is generally unsatisfactory. However, if a sensible legal view can be taken, then doubtless the position will be insurable. However, just as with title indemnity insurance, also restrictive covenant indemnity insurance, all insurance is a contract of indemnity and does not of itself remove a right. It is precisely because that right may yet exist and be enforceable that the insurance is taken out at all.

Easements are often found to be closely drafted, and thus they need interpretation in context. Sometimes the right is expressed in a conditional manner which has the effect of keeping the right alive whilst imposing a burden in the nature of a covenant, even though the person benefiting from the right did not enter into any such covenant as such. Thus, for example, a right of way may be expressed to be exercisable only upon the beneficiary either maintaining it or paying a contribution towards maintenance by others. Conversely, in general terms, a right of way carries with it an implication that the beneficiary is entitled to do what is necessary to keep it open including, for example, removing obstructions, repairing the surface and so on.

Rights of support and protection are particularly important, not least when buildings are dependent upon each other for maintenance of their structure but, say, one is to be replaced by a new development. The more confined the development area, the more one is likely to find that the scope for activity is confined by the rights of immediate, and not so immediate, neighbours. Where rights are expressed by deed, whether granted or reserved by a conveyance or transfer, or by a separate deed of easement, one should also not lose sight of other implied rights, and these may be far more problematical. At this point, the construction professional must be vigilant as he may be able to spot characteristics of a physical nature which it is not in the gift of a lawyer or indeed anyone else so untrained, as to recognise. Those characteristics, however, once identified may yet be the subject of legal rights, if not protected by statute then protected by common law. In the case of the latter, there may be legal easements as enforceable as those created by deed, but lacking only that characteristic.

A landowner who sells some of his land leaving his retained portion landlocked is thus afforded, at common law, what is known as a way of necessity. By contrast, if he sells the landlocked part first, it is not implied that the buyer has the right over the retained portion. For that, you must either have an express grant or exercise such right in an appropriate manner and for so long as to afford him a prescriptive right. The first signs to look for, therefore, are whether anyone appears to be exercising a right of whatever kind over the site. Perhaps it is the use of an old road or pathway. Perhaps there is a drain, the existence of which is not recorded in deeds. Perhaps there is a building close by, having windows in it, the light to which would be impaired by a new development, the list is endless and it behoves every construction professional to make no assumptions. It may be, on legal analysis, that the tables are entirely turned. For example, in the absence of express grant by deed, the neighbouring development might itself have been there for too short a period to acquire any kind of implied right and, if so, the owner will be powerless to prevent new development neighbouring.

The problem with implied rights is that unless established by agreement and reflected in subsequent deed, the claimant must move the court and establish his right by proceedings. He may, within those proceedings, seek and obtain an injunction. With that, of course, the house of cards created by a series of development agreements as discussed in earlier chapters may thus fail. However, say a building existed, or has windows in it which have existed, for less than a full period of 20 years. The owner of the would be servient tenement has the ability, under the Rights of Light Act 1959, to serve a light restriction notice which must then be lodged with the Lands Tribunal. Assuming it is not challenged, or at least not successfully challenged, the notice will eventually be registered as a local land charge and effectively block the acquisition of right of light by prescription, in the case of a right of light after a full period of 20 years.

It is possible to claim an implied easement at common law based on long enjoyment, difficult because the common law thinks in terms of 'time immemorial'. However, this is based upon legal theory which assumes a right going back to the first year of the reign of Richard I, 1189, and the adoption of 'lost modern grant', post-1189, is scarcely better particularly where buildings are concerned. Far better to rely upon the Prescription Act 1832, Section 2 of which provides (except in the case of an implied easement of light which relies on a 20-year period alone) that where an easement has been actually enjoyed without interruption for 20 years it shall not be defeated by proof that it commenced later than 1189, but it may be defeated in any other way possible at common law. It is then further enacted that an easement which has been enjoyed without interruption for 40 years shall be deemed 'absolute and indefeasible' unless it appears that it was enjoyed by some consent or agreement expressly given by deed or writing.

Interruption is problematical, in the case of statutory interruption unless it has been submitted to or acquiesced in by the dominant owner for one year after he had notice of it and of the person responsible for it. Interruption implies some overt act such as physical obstruction of a right of way. Rights of light consultants thus point occasionally to the idea of an easement persisting for 19 years and a fraction of year, which is largely academic and founded on case law. If the servient owner disputes a right before the expiry of 20 years, and succeeds, well and good, but if there is no interruption then the dominant owner can claim his right at any time after the full period of uninterrupted enjoyment of 20 years. It is most important to remember that implied rights must be uninterrupted, that is to say it is not a matter of the right being continuously exercised but that if there is exercise then that exercise itself must not be interrupted. Further, the right must be exercised as of right so that it is no use, for example, if there is an intervention where exercise is assisted e.g. by express licence. There may be other characteristics such as the servient owner, if that is what he is, not having the power to grant, and so on. Again, this begins to look a little academic but the key message for those involved in development is to be vigilant as to the rights of others and, where suspicions are aroused, for legal advice to be taken.

Implied easements are problematical. An easement of 'necessity' (land-owner's retained land being landlocked after sale of part) is confined by the actual use. An implied easement by 'prescription', needing to be claimed or agreed is also problematical as until an express grant the extent and scope will be circumscribed by the circumstances, backed by whatever evidence is available. Implied rights are further problematical in that whereas express rights granted by deed must suffer attack in interpretation to see how wide they are, implied rights are limited by the evidence. A right of way acquired by prescription may be of little use if its exercise throughout was confined by a use of the dominant tenement that itself has been abandoned. A wholly different use may thus be defeated.

Rights in leases should also be considered. It was mentioned, above, that Section 62 Law of Property Act 1925 had effect to transfer with land various rights associated with it, unless the contrary were found in the conveyance or transfer. The same applies to the grant of a lease and, indeed, occupational leases frequently expressly exclude such rights saying, in so many words, that only those rights which are expressly granted by the lease are granted to the tenant. So, in multi-occupation development, the rights conferred upon tenants (and indeed reserved out of the grant) tend to be confined to the relationship of the let premises to the greater development.

Finally, a word about the relationships between boundaries and easements. As has been seen (save as to licence – see below) exercise of rights which are not supported by deed can only be legitimised by prescription and, in practice, by the application of the provisions of the Prescription Act 1832. Moreover, even if the facts are supported by depositions perhaps only an order of the court, in the alternative, is foolproof. However, it is the careless way of the world that very frequently various rights are casually exercised and a view has to be taken. Expect the institutional commercial property market to be robust and to demand certainty.

Occasionally, what was perceived as institutional may come badly unstuck. And the same applies to all classes of property. The issue is one of the relationship of buildings to boundaries. So often one finds that, say, the flank wall of a building actually marks the boundary in question. Unfortunately, it is a natural characteristic of a building that it has footings which protrude unseen beneath the ground, eaves and gutters that overhang and, again, windows that open outwards. There is no basic right of law that permits any of these all of which, in the absence of grant by deed or prescription, are unlawful in the absence of licence. The concept of licence is essentially personal and whether or not it can bind/benefit successors will depend in law on factors beyond the scope of this book. However, true licence is, first, temporary in its nature but, second, and more importantly, it may constitute valuable evidence to rebut the acquisition of a right by prescription.

Accordingly, the designer who places the flank wall of a building precisely along the boundary of the site places the site owner in prospect of legal action to eradicate the associated incursion. An expensive negotiation may ensue (and should be pursued in case of accidental incursion) with the view to securing by that negotiation if not the grant of an express right then a licence benefiting successors for the duration of the life of the building. The fallback, unhappily, is capitulation to the rights of the dominant owner who, in any proceedings, will almost inevitably prevail. This chapter serves as a warning, therefore, to have regard to the property rights of others, whether or not expressed by deed.

# 6 Interpretation and some statutory hurdles

## Or: How to invite trouble?

The law is much easier to understand once one dismisses any notion that somehow we live and breathe solely by dint of statute. Our common law is judge-made law, tempered by equity. As to the latter, hundreds of years ago it was the king's Lord Chancellor who developed his own court to provide a remedy for wrongs which the common law was too narrow to address, resulting in injustice. The resulting principles are known as 'equity', allowing the courts of today the discretion to override the common law where strict application would result in an injustice. Today we see the successors to the courts of common law and equity, now blurred and distorted, in the Queen's Bench and Chancery divisions of the High Court. No longer, however, does one court administer a single set of doctrines or principles.

Layered onto this is statute, that is to say acts of parliament, and also statutory instruments, being secondary law drafted under powers contained in principal statute, all of which itself is interpreted by judges whose prerogative it is. (When some provision of a statutory instrument, usually called an order, is in issue, the validity or effect can thus be challenged in proceedings as being *ultra vires* the enabling statute.) Of course the doctrine of supremacy of Parliament is at once tainted by the supremacy of much European law to which we have elected to be subservient, and some judgements may appear perverse to government. However, it was government's choice to subscribe.

Legal precedent, that is to say case law, commences at High Court level, and is overridden by the Court of Appeal and ultimately by the House of Lords. What is not challenged by higher authority lies where it falls for the time being. That is not to say that decisions of the County Court are necessarily to be ignored. Sometimes County Court decisions are reported as highlighting a particular point of interest or difficulty, and first-instance construction proceedings are frequently reported. Whilst precedent is not created, decisions of the lower courts may still be helpful in developing an argument or principle, as subscribers to construction law periodicals will know.

This blend of measures is broadly termed the common law system. It is, therefore, to be distinguished from codified systems. It also means that whatever the matter in issue, a broad approach has to be taken to interpretation.

A good lawyer is someone who does not readily jump to conclusions and the layman should not be afraid of statute. The best rule of thumb is to read the words according to their common meaning, except where a word has been afforded a statutory meaning in context. Sometimes a term is given meaning by the section in question. Other times, an interpretation section may be found.

People concerned with development find themselves hedged about by statutory inhibition of all kinds and then, when a transaction finally emerges as a viable proposition, its conversion into contractual form, including security of obligations, is necessarily subject to formal legal documentation. There is much the non-lawyer can do for himself when addressing a legal document, the first is to use common sense and then attribute to words their ordinary meaning. The second, a little more lawyer-like, is not to assume inconsistency between one provision and another. If pressed, the answer to the apparent inconsistency is that a later inconsistent provision may prevail over an earlier one, unless the contrary be clearly shown.

Those who have come across ancient conveyancing documents will have seen how some draftsmen would use half a dozen or even more consecutive words, each having a similar meaning, in order to ensure that a single notion was clearly expressed. Some find the absence of punctuation also causes difficulty. When drafting legal documents, the lawyer is attempting to convey a single meaning. If that is achieved then punctuation becomes unnecessary. As we all know, insertion of punctuation at different places in the same set of words can have the effect of altering the meaning, perhaps conveying the complete opposite. Good legal drafting is not the prerogative of modern draftsmen, however, and the same applies to statutes. One only has to look at older statutes, not to speak of commercial agreements, to appreciate the clarity with which they are expressed. Today's legislation leaves much to be desired, so also does the quality of much commercial drafting.

Furthermore, whereas once upon a time governments legislated in fairly small quantities, in today's world statutes and statutory instruments, often European-driven, proliferate beyond the absorption rate of most mortals. For any construction professional reading this book, his own world provides proof of that, but the prime purpose of this book remains to broaden that perception into an appreciation of development and the law at large. The following topics, therefore, are a collection, hardly random, and still less complete, but they illustrate the application of certain aspects of statute law to the development process. This chapter flags up some ideas, and for a detailed explanation one should turn to specialist works and legal advice as appropriate.

## Landlord and tenant

For most construction professionals, landlord and tenant issues are largely a matter for the employer alone. Occasionally, however, where site assembly depends upon termination of tenancies, a start on site may be unexpectedly delayed. However unmeritorious his case, a tenant of business premises, say, might use his lease as a delaying tactic for obtaining possession, possibly even where the tenancy was contracted out of the security of tenure provisions of the Landlord and Tenant Act 1954 by the obtaining of a consent order of the court at the outset. Usually, the presence of such an order leaves little doubt, and the worst that may usually be expected is proceedings for possession.

Where the tenancy is protected, however, ground (f) of section 30 of the 1954 Act, an intention to demolish or reconstruct, can be problematical. There is a line of cases highlighting aspects of the issue, including that the intention is not necessarily undermined by the absence of planning permission as at the date of the hearing. However, timing of commencement of the development may be impaired if the credibility of the case in support of the particular ground for possession is vulnerable to attack and may thus have a direct impact on the programme. Only when vacant possession can be given should the programme start date be fixed and developers who ignore occupiers' rights do so at their peril. If anyone is apparently in occupation as the start time approaches, one should question it.

The Landlord and Tenant Act 1954 is thus an extremely important statute. It is peculiar to England and Wales and does not apply to Scotland. In affording security of tenure it thus also presents problems in site assembly. Part 2 applies to premises which are used for the purpose of a business and, as one might expect, there has been a raft of cases over time as to what constitutes a business for the purposes of the Act, not least when premises contain an element of residential occupation. These issues apart, the key characteristic of a business tenancy protected by the act, is that the contractual term itself continues beyond the term defined in the lease or tenancy agreement until it is brought to an end by one of the means prescribed by the act. It works both ways, however, in that although the tenant under such a tenancy can walk away at the end of his original contractual term, if he stays any longer with the acquiescence of his landlord he must serve on the landlord what is known as a Section 27 notice giving not less than three months' notice expiring on a usual quarter day (24 March, 24 June, 29 September or (yes, Christmas Day) 25 December).

A landlord serving a notice to terminate the tenancy must serve such a notice in a statutory form (a Section 25 notice) to expire not less than six months nor more than twelve months after the giving of the notice, whereupon the tenancy will come to an end. Moreover, the landlord must show at least one of a number of statutory grounds set out in the act if he wishes to obtain possession. He might, of course, be willing for the tenant to have a new tenancy and should therefore so state in his notice. A tenant prospectively to be ousted by such a procedure must respond within two months saying whether or not he is willing to give up possession. Moreover, if he is not willing to give up possession, he must make an application to the court by the expiry of the fourth month, otherwise he loses his statutory protection.

The landlord's statutory notice is given under Section 25 of the act, a tenant's statutory notice, if he wishes to apply for a new tenancy, is prescribed under Section 26 of the act, and once he supported this by an application to the court as before, otherwise his rights were similarly lost. Since June 2004 the law has changed (see below). It is not entirely surprising, therefore, that there is much case law surrounding these statutory provisions, and there are plenty of hurdles for a landlord involved in site assembly whose solicitors fail to ensure that procedures are effectively followed and tenants deprived of the opportunity of making mischief. If a tenant has merely bought some time in the process, he may yet have a bargaining position as a basis on which to agree to surrender his tenancy at a certain time. However, just as a business tenancy could be contracted out of the protection of the act by the obtaining by both the landlord and tenant of a consent order of the court, an agreement to surrender a protected tenancy was void at law unless a consent order was also obtained. All of these court proceedings have been replaced by notice procedures.

Whilst on the subject of landlord and tenant, it is perhaps helpful to expand in the context of investment. As has already been seen, the capital value of an investment reposes not in the bricks and mortar of a development, but in the income stream from tenancies. Investment surveyors will concern themselves, therefore, not only with commercial terms of leases but also with covenant strength. The Landlord & Tenant (Covenants) Act 1995 brought about an important shift effective for new tenancies created after that year. The doctrine of privity, as it was known, ensured that unless expressly released from his obligations, an original tenant under a lease remained liable for the obligations of the tenant for the time being under the lease. Moreover, although the law was also, and remains, that, after the original tenant, the tenant for the time being remains liable only for the period of his tenure, so that he ceases to be liable upon assignment, in those cases where the landlord's consent to assign is required it was, and remains, the practice that in any licence to assign, the landlord would take a covenant from the prospective assignee undertaking the obligations of the tenant for the entire residue of the term. Thus, as leases changed hands, landlords would have a pool of former tenants, underpinned by the original tenant, on whom to draw in case of default on the part of the tenant for the time being.

However, for all new tenancies created after 1995, the law has changed significantly. By 'new' one must exclude tenancies arising by way of statutory entitlement to renewal or an option to renew contained in a lease granted before 1996. Now, a tenant is liable only during the period that he holds the lease, but his landlord is entitled to require an original tenant or his successor to enter into what is known as an 'authorised guarantee agreement' which effectively covers the period of tenure of his immediate assignee. The liability therefore drops away upon assignment and the process can be repeated. The landlord's entitlement, whilst facilitated by statute, must nonetheless be incorporated as a term of the lease, without which the landlord cannot avail himself of it. Some leases will frame the entitlement on the basis that 'the landlord reasonably requires', but this is clearly open to debate and invites dispute. A lease is better drawn on the basis that the landlord requires in any event.

Of course, these changes brought pressure upon the concept of the landlord's consent 'such consent not to be unreasonably withheld', with the consequence that there was a tendency in post-1995 leases to add qualifying criteria designed to ensure maintenance of covenant strength if a tenant were minded to dispose. There is, however, the danger of onerous obligations impacting on rent on review. For example, if a lease precludes assignment or underletting absolutely, it is only of value to the original tenant and is inherently unmarketable.

As time has gone by, however, and practitioners have become more used to the impact of the act, and apart from the express imposition of a landlord's entitlement to require an authorised guarantee agreement, alienation clauses are beginning to look much like their pre-1996 forebears.

If an assignee goes into liquidation, his predecessor (whether or not he gave an authorised guarantee agreement) can be required to make payments to the landlord for the entirety of the residue of the term as if the lease still existed, even though there is no tenancy. This liability arises under Section 17 of the 1995 Act applies to both older and new tenancies. In those circumstances the predecessor might better be advised to exercise rights subsisting under the 1954 Act and have the lease vested in him. At least he will then be in some control of events but not entirely as he then needs to occupy or dispose. The landlord may apply for such payments by serving notices under Section 17 of the 1995 Act. However, this is not entirely satisfactory for him either. If he wants to see the premises occupied, and not continuing a discouraging void in the middle of, say, his shopping centre, he needs to be proactive and seek to re-let as soon as possible. There are thus conflicting interests and pressures.

Changes in conveyancing practice cannot pass without the addition of a comment on landlord and tenant procedures. For a business tenancy to be excluded from the effect of the security of tenure provisions afforded by the Landlord & Tenant Act 1954 Part II, the procedure has hitherto been for the parties to the tenancy to seek a consent order of the court. Similarly, whilst a surrender of the lease might be perfectly in order, an agreement to do so was void without a like order being obtained. New statutory provisions, effective 8 June 2004, provide a mechanism circumscribed by the giving of prescribed notices, compliance with which obviates the need for a court order.

## Planning and highways

No, this is not an attempt at reduction of planning law to a few handy pages. The issue is viability and implementation. By the time a planning permission is challenged, progress has transferred from its initiators, an architect, planning consultant or whoever, to the lawyers.

First, the concept of judicial review is not a product of planning legislation. It is a creature of court practice rules which allow anyone with an interest, and not an applicant or an owner, to challenge the propriety of the grant. It used to be the case that an applicant for planning permission, having achieved his desired grant, could reasonably safely regard a period of three months and one day as effectively preserving the permission he had obtained. Conversely, a third party aggrieved person should allow no undue delay, and possibly even as little as six weeks from grant might itself be unsafe. Just as further appeal by an applicant on a point of law from an appeal to an inspector (see later in this section) is confined to a point of law so is it also with judicial review. An aggrieved person cannot challenge the decision but can challenge the process and it is 'due process' that has been the subject of a number of cases which make it difficult to take a view as to whether a planning decision is implementable.

In *R. (Gavin)* v. *Haringey London Borough Council* the issue before the court was of apparent failure by the local planning authority adequately to publicise a planning application. Mr Gavin (a solicitor) lived across the road from the site and was not notified of the application although others had been. No site notice had been erected. Planning permission was, in fact, granted in 2000 and, 32 months later in March 2003, the first Mr Gavin knew about it was when work started on the site opposite. That he was permitted to proceed at all is remarkable. There had been such a delay, however, that on this occasion the court did not quash the planning permission. The court determined that the local planning authority had indeed failed to comply with publicity requirements and, indeed, also environmental impact assessment obligations. There had been serious breaches but the developer was entitled to rely upon the local planning authority discharging its obligations and was not obliged to monitor the steps taken to comply with them. Third parties could rely on information in the planning register and quashing a planning permission so long after it had been made would not only be detrimental to good administration but also cause hardship and prejudice to the developer. What is to be made of that?

First, where there is a failure of due process, delay in bringing a case may simply arise through that failure being unascertainable. Suppose, in the case of Gavin, that the period had been rather shorter: precisely when, had the developer commenced work earlier, might it had been quashed? The answer, unhelpfully, is that cases turn on their facts but the emerging principle is that bland assumptions about the validity of a planning permission cannot be made. Further, a developer cannot wait indefinitely on the off-chance that his planning permission may be challenged. In the Gavin case, a challenge arose as a direct result of visible commencement of work.

Neither is it a sensible proposition to suggest that one must stand behind planning officers and watch them at work. Applicants for planning permission need to attempt a balanced and practical view and, if the local planning authority appears at any stage to have taken a shortcut or omitted to do something which it should have done, that is nothing to rejoice over: applicants should be as vigilant as they can sensibly be to see that due process is observed. So, if an applicant is as satisfied as he can be, how long following grant should he wait before implementing his permission? An emerging view seems to be that one should probably now wait up to six months on the understanding that very occasionally, such as with the Gavin case, failure of due process may yet emerge. Nothing is perfect.

The above should not be confused with an application to the court following an unsatisfactory appeal or calling in. First, it should be remembered that in cases of non-determination there is a two-month back-stop, the expiry of which affords the ability to appeal on grounds of non-determination. Appeal is available (once again) within six months of actual or deemed refusal. Once the appeal has been determined, appeal to the court, that is to say the High Court, is available on a point of law if an application is made within six weeks of the appeal result. In the case of judicial review, mentioned above, non-compliance with statutory requirements may extend to breach of the Human Rights Act 1998. This time, however, the ability to appeal is confined to the original appellant or a person with 'sufficient interest', such as his successors in title. An appeal on a point of law is thus not capable of embracing the appeal result at large. This is in the sense that the planning process is primarily in the hands of the local planning authority and, if required, an appointed inspector. Moreover, the emphasis today is on plan-led development. Section 54A of the Town & Country Planning Act 1990 says 'Where in making any determination under the Planning Acts regard is to be had to the development plan the determination shall be made in accordance with the plan, unless material considerations indicate otherwise'. The planning authority must not simply have regard to the plan, but decisions must now be made in accordance with it except where 'material considerations indicate otherwise'. An appeal on a point of law will thus be essentially on the basis that either the Secretary of State's decision was *ultra vires* the Town & Country Planning Act 1990 or that statutory requirements were not complied with.

Another important area is statutory agreements, with the subject matter of which professionals will largely be familiar. Whether required or proper is another matter. First, a condition in the planning permission itself may be held to be void if it relates to a subject matter outside the application site. The concept of 'planning gain' is wide indeed. Moreover, a mere condition, whether of construction or use of the built outcome may insufficiently express or satisfy the requirements of the planning authority. A statutory agreement under section 106 of the Town & Country Planning Act 1990 fits the bill as being binding on the person entering into it (who must have sufficient interest in the land the subject of the (intended) planning permission) and on all successors. And in case the solicitor has forgotten to make such a provision, always check such agreements for a provision that the burden of the agreement will cease to bind a person once he has parted with all his interest in the relevant land, and further that he should not become bound in any event until the related development commences. Some such agreements sometimes impose conditions whether or not the development commences, and are the rationale for the agreement being entered into at all.

The fact that one has secured a planning permission does not, however, mean that one is entitled to enter into a Section 106 agreement, save to subscribe to obligations, which on its own does not suffice. A planning application can, of course, be made by anyone and provision is made in the legislation for those who have actual interests in the land to be given formal notice of the application. Suppose, however, that the resolution to grant or a condition in a planning application itself requires a Section 106 agreement to be entered into. Unless the applicant has sufficient interest he cannot enter into the agreement, so he must look to someone sufficiently interested, i.e. the landowner in person. Nothing can compel the landowner to enter into such an agreement (unless of course he has agreed to do so, say, under the terms of a sale and purchase contract). It is worth mentioning, therefore, that a contracting purchaser does not have sufficient 'interest in the land' to bind the land for the purposes of a Section 106 agreement. Similar considerations apply in relation to other forms of statutory agreement such as those under Sections 38 and 278 of the Highways Act 1980 or under Sections 98 and 104 of the Water Industry Act 1991.

It is established that when a statutory agreement derives from a condition, then that condition should not itself be used as a device to thwart implementation of the planning permission. At once the notion of reasonableness is injected. However, beware generalisations, analyse the circumstances and take legal advice. It is an area open to abuse.

The object of this short section on planning was to highlight some of the issues which consistently require attention. The subject cannot pass entirely without mentioning human rights, and the impact of the Human Rights Act 1998 and the European Convention on Human Rights. Article 6 entitles everyone, in the determination of civil rights and obligations to 'a fair and public hearing ... by an independent and impartial tribunal ...'. A number of cases have centred on two other provisions in particular. Article 1 of the first protocol, protection of property, states 'Every natural or legal person is entitled to the peaceful enjoyment of his possessions. No-one shall be deprived of his possessions except in the public interest and subject to the conditions provided for by law and by the general principles of international law'. Article 8, as to the right to respect for private and family life, states in particular that, 'Everyone has the right to respect for his private and family life, his home and his correspondence'. It goes on 'There shall be no interference by a public authority with the exercise of this right except such as is in accordance with the law and is necessary in a democratic society in the interests of national security, public safety or the economic well being of the country, for the prevention of disorder or crime, for the protection of health or morals, or for the protection of the rights and freedoms of others'. Article 8 is thus concerned with preventing intrusions into someone's private life, particularly arbitrary intrusion.

It is also unlawful for a public authority to act in a way that is incompatible with a convention right, but it will not be acting unlawfully if it could not have acted in any other way because of one or more provisions of primary legislation. If the legislation itself is incompatible it is not invalidated. Rather, the minister of the crown in question is empowered following a declaration of incompatibility made by the court to order amendments to be made to the legislation. Of one thing there can be no doubt at all: techniques of appeal and challenge are forever evolving; further leading cases will inevitably emerge.

Neither can these thoughts on planning be allowed to pass without mentioning, very briefly, environmental impact assessment. The current regulations are those encapsulated in the Town & Country Planning (Environmental Impact Assessment) (England & Wales) Regulations 1999. Once upon a time, these regulations and their forebears of 1988 were perhaps not at the top of people's lists of concerns. Lack of attention to detail can be the downfall of an application. However, cases on the subject are developing and it would be wrong to assume that simply because an application was in outline only it would automatically fail. The answer is, and the regulations allow for it, that close consultation with the local planning authority is required. The subject having been flagged, it is beyond the scope of this book to expand upon it.

Finally, this book has gone to press too late to comment on the Planning and Compulsory Purchase Act 2004. Suffice it to say that the principles of plan-led development have been re-shaped entirely, and a specialist work should be referred to for the fundamental changes that were introduced from 28 September 2004.

## Other government intervention

If influence over the development process were confined to interference by local planning authorities working within the framework of planning legislation, in turn steered by planning policy guidance (PPG) notes, and the ultimate intervention of courts, the process would be 'relatively' simple. From the development corporations of twenty years or so ago and then enterprise zones (almost dead, but not quite as has been seen) there has been a proliferation of statutory 'corporations' usually described as an 'agency'. In the eyes of successive governments, such corporations have replaced branches of the civil service in the interest of some kind of efficiency. Public and quasi-public sector players appeared as a heading in Chapter 1, so now for a little detail.

Such bodies are corporate entities: they are not companies, they have no shareholders, they are funded by (and at the whim of) central government (although the supposed efficiency sometimes manifests itself in the making of charges, thus removing some of the cost of administration from ordinary taxation), and are accountable to the appropriate Secretary of State. As corporations, however (but not limited companies), they are legal entities and so can acquire, hold and dispose of assets, enter into contracts, give security and so on. Being creatures of statute, they run the risk of being tainted by *ultra vires* (see as to NHS trusts in Chapter 1) but there is a thread to most of their enabling legislation. Perhaps the greatest error of assumption, however, would be that because such a body is created and controlled by central government, somehow its obligations are guaranteed. They are not. However, running through the enabling legislation for such bodies is one notion in particular that anyone dealing with such a body, say acquiring an asset from it, shall not be required to 'see or enquire' in so many words as to the propriety or validity of the dealing. However, that is not quite the same as saying something like 'but if I do find something out, perhaps I can ignore it'. Company lawyers, for example, will point to the possibly limited value of protection of transactions authorised by board resolutions, under Section 35 of the Companies Act 1985. Given in particular that the enabling legislation for such statutory corporations also contains express prohibitions, qualifications and so on, and just as we have seen in the past issues of *ultra vires* coming before the courts in connection with local government, there is plenty of scope for keeping lawyers busy in relation to other public-sector bodies. Even local government has had to be revisited in terms of the Local Government (Contracts) Act 1997 which was enacted specifically to validate contracts with a PFI element and whose relatively short provisions should be noted with care by the lawyers concerned.

## The utilities

Gas, electricity, water and sewerage have always seemed to be a bit of a mystery. Remembering that the post office and telecommunications were once also combined and that all of the undertakings were publicly owned, we have come a long way in the past 20 years. For those involved in procurement of development, securing of co-operation to bring in essential services is essentially a matter of agreement. Where pipes, wires and cables are to be laid, that agreement will translate itself into deeds of grant on agreed terms, in the case of overhead electricity lines, telegraph poles being the subject of a wayleave, which is a form of licence. So far so good.

The fact remains, however, that even with today's privatised utilities, indeed public companies whose shares are traded on the London Stock Exchange, the public services which they perform are heavily circumscribed by statute. This covers not only what they do and how they do it but also, very particularly, whether they may or should do it.

All development prospectively requires the co-operation of the utilities to provide electricity, water and gas. It would be wholly wrong to assume that these facilities can be demanded and obtained at will. In all cases there may be grounds for non-provision, backed by the authority of statute. Of the three, perhaps water is the most problematical, given water companies' expressed fears that resources may not suffice.

The gas industry came under British Gas PLC pursuant to the Gas Act of 1986. There has been further fragmentation since. A public gas supplier's duty is, so far as it is economical to do so, to provide a supply upon reasonable request. In case of the shifting sands of the industry, it is perhaps unwise to think in terms of 'thresholds' but an obligation to supply only rises if the premises are within 23 metres of a distribution main where the supply will be less than 732,000 kilowatt hours.

The utilities are characterised by each having a director-general to whom applications are made and who also has certain specific powers. In the case of gas these include compulsory acquisition powers in support of the authority's functions. These include rights to store gas in an underground gas chamber.

In turn, electricity was privatised under the Electricity Act 1989, with similar characteristics including compulsory powers. Electricity has its own characteristics, of course, and so there is a statutory duty, for example, to take all reasonable precautions not to interfere with the operation of telecommunications apparatus under the control of a licensed provider of telecommunication systems. Again there is a duty to supply, but there are exceptions including a broad sweeper that it is not reasonable in all the circumstances for the supplier to be required to do so. Failure to supply when a supply should be provided may lead to an order by the Director-General of Electricity Supply but only in case of breach of an order will an aggrieved person have a remedy (now civil, although once upon a time penalties could be imposed).

The water industry in particular should be noted, not least because of the environmental aspects of its statutory controls. The industry was privatised under the Water Act 1989, the process being developed by the Water Industry Act 1991. First the National Rivers Authority and, following the Environment Act 1995, the Environment Agency became the governing body of the water companies which actually provide the supply. The Secretary of State and the Director-General of Water Services have a duty to secure that the functions of the water undertakers are properly carried out.

As with other utilities there are duties to connect to a supply, but again hedged with qualifications. Beware reading the subject too lightly: reference to domestic supplies actually includes supplies to commercial property. In some ways analogous to the other utilities, a civil remedy exists for failure to supply water but it is a defence that the undertaker took all reasonable steps and exercised all due diligence to avoid breach of that obligation. Note particularly, however, that there is no duty to provide water for domestic purposes unless the undertakers cannot meet all the obligations without incurring unreasonable expenditure, or if in so doing they put at risk their obligations to other domestic users.

As with the other privatised industries, an undertaker or the Environment Agency itself may be authorised to acquire land or rights compulsorily. A compulsory works order may itself necessitate to the acquisition of landlord rights.

A number of changes have been brought about following the Utilities Act 2000 which created a combined gas and electricity markets authority. The Water Act 2003, which received Royal Assent on 20 November 2003, makes a number of further changes including a new regulatory authority, the Water Services Regulation Authority, replacing the Director-General of Water Services. The Act deals with a number of miscellaneous provisions and, if one must cherry pick in the development context, it will be noted that there are provisions enabling a developer to enter into an agreement with a water undertaker to lay water mains and communication pipes in accordance with standards set by that undertaker, and also for the Secretary of State to be empowered to make regulations to place a duty on sewerage undertakers pursuant to their existing powers under Section 102 of the Water Industry Act 1991 to adopt private sewers in defined circumstances. A number of broader provisions place duties on public authorities as to conservation of water.

## Contaminated land

The raft of legislation since the 1980s has transformed the development scene. The presence of contamination has created a breeding ground for specialist consultants concerning whom one is warned, from time to time, to question their qualifications or experience. However, with the passage of time, the importance of the subject has given rise to the emergence of consultancy firms who have also had time to gain a reputation and a track record. The related legislation is formidable, but not intellectually unassailable, albeit legal practitioners professing expertise in the area will tend to specialise.

The impact of environmental law now goes to the root of the whole of the development process. One does not have to be developing a brown-field site. In 2001 the Law Society issued a green warning card to solicitors which includes the following advice:

> The advice contained on this card is not intended to be a professional requirement for solicitors. Solicitors should be aware of the requirements of Part II of the Environmental Protection Act 1990, but they themselves cannot provide their clients with conclusive answers. They must exercise their professional judgement to determine the applicability of this advice to each matter in which they are involved and, where necessary, they should suggest to the client obtaining specialist advice. In the view of the Law Society the advice contained in this card conforms to current best practice.
>
> Solicitors should be aware that environmental liabilities may arise and consider what further enquiries and specialist assistance the client should be advised to obtain.
>
> In every transaction you must consider whether contamination is an issue.

It would be extremely unwise to attempt here to condense the law. The next section in this chapter, 'Nuisance', is a natural bed-fellow of environmental law. Long before the current raft of legislation the common law, through the tort of nuisance, provided effective remedies. However, one needed to be an aggrieved party, one who had suffered damage in the common law sense, in order to effect a meaningful remedy. The imposition of statute, however, adds a new dimension by imposing regulation. The mainstay today is the Environmental Protection Act 1990 all of which was in force by 2000 and which commences by defining contaminated land. (The Water Act 2003 also makes changes to the definition of contamination land in relation to water pollution. These represent a relaxation of requirements concerning minor cases of water pollution.) It is helpful to remember that however much land is contaminated, if the contamination is wholly contained so that it is harming no-one, it is not necessarily perceived as a problem. If it threatens to harm people or property, water supplies or wildlife that is another matter. Subject to this, local authorities (and, indeed, also the Environment Agency) have special duties to deal with it. Following notice to remediate, the Environment Agency, in particular, has power to prosecute in case of water pollution. The essential thing for all involved in development is to be on enquiry, and solicitors, in particular, may also be expected to review their local land charges searches in this connection which may throw up a number of things, for example, if there is a statutory nuisance (see under 'Nuisance' below), that the landowner has failed to comply with an abatement notice, or where there is the designation of a smoke control area, and so on. Enquiries of the Environment Agency may yield information on contamination, and in particular the existence of former landfill in the vicinity, and whether the property lies within a 100-year flood plain. Enquiries should also be made by solicitors of sewerage undertakers and water companies and, as regards nature conservation matters, English Nature. Mention should be made of the Environmental Information Regulations 1992 which follow the EC directive on the Freedom of Access for Information on the Environment. Some familiarity with these regulations may prompt more searching enquiries of various authorities. There is no lack of useful information, but all too often there is a failure to seek it.

Arising out of the 1990 act are the Pollution Prevention and Control (English & Wales) Regulations 2000 which deal with such matters as designation of special sites, pollution of controlled waters (itself a technical term) and the general fabric of serving and implementing remediation notices.

Slightly further removed are nature conservation issues dealt with under the Wildlife and Countryside Act 1981 and, inevitably, further European regulation in the form of the Birds Directive and the Habitats Directive.

Where contamination requires removal of soil, it may be removed and sent to a licensed infill. Landfill tax was imposed under the Finance Act 1996 subject to certain exemptions (for which an information note is available from Customs & Excise). Certain tax reliefs are also available against the cost of cleaning contaminated sites, and developers should explore these aspects with their tax advisers.

From the lawyer's perspective, whoever he is acting for in the development process, it is not just a matter of identifying environmental issues, evaluating them in context and ensuring that the need for technical advice is brought to his client's attention when necessary. Crucially, his role must be seen as one of avoidance, a little bit like an accountant trying to shorten the tax bill for his client. The legislation does not actually say that 'the polluter pays', rather that liability falls on 'causers' and 'knowing permitters', and if none are found, then on owners and occupiers. In the case of water pollution statutory liabilities fall only on causers and knowing permitters but at common law, common law liability falls on those who knew or ought to have known of the pollution of whatever kind to the extent that they fail to take whatever steps were reasonable in all the circumstances. Put another way, it is simply a manifestation of the common law doctrine of the duty of care which we all owe to each other, whatever the issue or circumstances.

If a seller of land wishes to divest himself of liability he will fail to do so unless the buyer is afforded sufficient knowledge of the contamination in question. Liability can be transferred to the buyer by agreement, but the buyer must be sufficiently aware of the 'broad measure' of the contamination. A typical clause will identify contamination by reference to a broad description (which may be highly speculative, but no matter) and then go on to acknowledge that the buyer has had the opportunity to carry out its own inspections and surveys etc., that it has satisfied itself and that the price to be paid for the land reflects all such matters. However skilful the wording, it will be entirely undone if there is a failure to disclose, the more so if there is a deliberate withholding of information which the buyer could not possibly find out for himself. Solicitors acting for buyers and their funders, tenants and so on need not only to frame their enquiries with care but those acting for sellers must be even more careful in their replies. For example, a reply to an enquiry in terms of 'not so far as the seller is aware' raises the spectre that, in order to provide that answer, the seller must have placed himself on a certain measure of enquiry. That measure may prove a subject of dispute or a claim for misrepresentation.

A seller should be no less nor more wary in replying to environmental enquiries than he is when answering others. To pass on such liability as he may have, disclosure is essential. Disclosure itself raises legal issues of its own: fortunately, insofar as the information could be ascertained by the buyer on making his own enquiries, then the inadequacy of the reply, if inadequate it appears, may be made good by the seller making clear that, notwithstanding, the buyer must place reliance on his own enquiries. Safety margins are not absolute, however, albeit that it is clear that negligent replies and dishonest replies will have well-deserved consequences.

Sellers with old reports, even recent ones for that matter, and other information which they have obtained from third parties may provide that information. The fact that they may themselves have relied on it does not mean that they should require buyers to rely on it, indeed quite the contrary. Professional reports are often written expressly on a personal basis for the very reason that the 'who is my neighbour?' principle may otherwise apply.

Moreover, the provision of such a report, even if it does contain such a qualification, may impute to the seller his own authority for it. He must take care to provide it expressly on a without responsibility basis. Carelessness by solicitors, whether acting for buyers or sellers can only provide work for others. There is, however, one curious phenomenon which perhaps this book may succeed, in some small way, in promoting: it comes down, again, to developers being encouraged to have their professional team and solicitors work together, rather than penny pinch and instruct on a need-to-know basis. Construction professionals and contractors involved with ground work, in particular, should be encouraged to call for solicitors' reports as well, and review them against their own enquiries and findings. Remembering that solicitors are lawyers and not scientists, the imagination which they have displayed may soon be brought to light.

Lastly, and particularly with the benefit of combined professional input, circumstances may arise when it is appropriate for a buyer to position his seller by requiring from the seller a warranty, for example, that this or that activity has not taken place on the site (particularly where contamination is known and disturbance, or the prospect of disturbance, is potentially problematical). Conversely, where uncertainty exists as to the contractual allocation of liability, a buyer's indemnity may be helpful. The more specific the indemnity, in case of known contamination, the better. But, where there is uncertainty, the wider also the better. As between sellers and buyers, and buyers and their funders, joint venture partners and tenants and so on, contaminated land issues are there to be managed, and not to overwhelm them. It is unhappily the case for solicitors that, as with everything else with which they are concerned, once in contact with an issue they carry the burden of it unless excluded from responsibility, either expressly or because the client can be shown to have sufficient understanding. Arcane reports and advice may result, but enquiries and replies are critical, the content of contracts the more so.

## Nuisance

Environmental law and nuisance are topics which naturally overlap. The leaching of contaminants may amount to a legal nuisance as well, so that modern statute is hardly a panacea for environmental remedy. The common law is centuries older. Moreover, other things can cause a nuisance, not least noise. Noise emanating from buildings can constitute a nuisance at common law, and also by virtue of statute. Nuisance is essentially interference with a person's enjoyment of rights over land, that is to say, the use or enjoyment of that land. The principle in the 1868 case of *Rylands* v. *Fletcher* showed that anyone who keeps anything on his land which is likely to do mischief if it escapes, keeps it at his peril. The damage, in order to be actionable, must be foreseeable so that the principle does not apply to things which are naturally on the land, or which escape in consequence of an act of God, or of a stranger, or the default of the claimant himself, or with his consent. Needless to say, the principle itself is very broad and, whether a nuisance has been caused or not, may rely heavily on the evidence in context.

Much has been said in Chapter 5 about the rights of third parties over and against the land comprised in the development site. It is not just easements and general rights in land which may be offended. Moreover, the fact that planning permission exists is no indication that other legal rights may not be offended. Nevertheless, the proposed use and enjoyment of the resulting development may create other kinds of legal difficulty, not least that of nuisance which may survive more bureaucratic intervention to emerge as a statutory right directly enforceable by some member of the public or other 'legal person' (which of course includes a company or statutory corporation). The fact that the noise abatement officer of a local authority takes the view that a particular activity will not cause a nuisance neither deprives a person of his remedy at common law, nor indeed the local authority itself in relation to statutory nuisance. The whole notion of statutory nuisance relies on its common law counterpart (concerning which it is no defence that the complainant came to the nuisance). There have been a number of cases well publicised in the national press showing that nuisance, in lower case, is not necessarily a nuisance in law. Farms and farm animals, for example, create noise and dirt, quite naturally. So the first thing to do when anyone complains of a 'nuisance' is to test that proposition in its own right, and if necessary take legal advice.

Section 79(1)(g) of the Environmental Protection Act 1990 prohibits noise that is 'prejudicial to health or a nuisance'. Where a local authority is satisfied that a statutory nuisance exists, it must serve an abatement notice on the purported offender which is appealable in a magistrates' court. However, proceedings can also be brought in the magistrates' court and where the offence is committed on industrial trade or business premises, for example, on summary conviction a fine not exceeding £20,000 may be imposed. An available defence may include proof that the best practical means were used to prevent or to counteract the effects of the nuisance. One can also call in aid of a defence to a statutory nuisance that one was within the confines of a noise reduction notice under Section 66 of the Control of Pollution Act 1974. 'A noise reduction notice may specify particular times, or particular days, during which the noise level is to be reduced, and may require the noise level to be reduced to different levels for different times or days.' There are also provisions in the 1974 Act regarding noise abatement zones, a register of noise levels and related offences. A consent given by the local authority under that act to exceed the noise permitted and a noise level register does not constitute any ground of defence to proceedings brought by aggrieved persons concerning statutory nuisances. It should also be remembered that a remedy for statutory nuisance is not confined to the local authority itself. In short, whilst there may be sensible comfort in assurances comprised in correspondence with the local authority, and whilst there may even be some evidential value, statutory nuisance and common law nuisance remedies are not necessarily impaired.

# 7 Construction procurement in the development process

## Or: Before you build a house of cards, first get a pack

This chapter is not a beginner's guide to quantity surveying or an insight into its arcane practices, or indeed a commentary on the practicalities of procurement which are the domain of construction professionals.

Neither is it a digest of construction law, or an attempt to usurp the proper role of the construction professional in procurement. It is in substance a comment on legal contractual practice and its evolution and, most importantly, given the main thrust of this book, its relationship with the development process at large, as opposed to construction, and ways in which this is contractually expressed. Much construction, hedged around with the raft of planning, environmental, and health and safety regulation, and so on, simply culminates in the process itself, preceded by the appointment of professionals, the letting of one or more building contracts and related sub-contracts, the procuration of the works, payment and, ultimately one hopes, client satisfaction.

From the perspective of the developer, as fashioned and contrived by his lawyer, construction procurement is much more besides, as an element of a wider picture. It interfaces with other contractual relationships, for example those described in Chapters 2 and 3, it contemplates the possibilities and prospects of failure in part or in whole of the contractual relationships created by related development agreements, and it also looks as it must to the consequences of success or failure. The legal process in any commercial contractual scenario is as much as anything about fallbacks, the discouragement of recriminations and the accommodation of all the players in the piece some of whom, perhaps but for this book, are unseen. (The difficulty is that the last few comments will even be novel to some, but these propositions are nonetheless precisely how it is and the lawyer fails to address them at his client's peril. The cavalier or foolish client sees little point in it all, unless he reads this book, and if he does not end up bankrupt, he can be sure not to rise high in the industry.)

Any default, however arising, and perhaps far removed from the construction process, may lead to disposal of the land, it may be also lead to transfer (or 'assignment') of the benefit of related construction contracts to third parties, whether or not the development is complete. It may, in the extreme, lead to abandonment, recrimination, and the discovery by the professional team that everyone has a fallback of some kind, except, unhappily, them. They are the sacrificial lambs, and Chapter 8 will seek to show how things start to unravel, how important it is for the letter of contractual relations to be addressed, and some of the practical consequences. Hence its subtitle 'fallback in action'. The quantity surveyor reading this chapter may thus breathe a sigh of relief that his domain has not been encroached upon or his role usurped, but he will perhaps now sit up and take note of how his fate is prescribed, far beyond the bounds of the document in his hands, his appointment. How did that arrive?

In Chapter 1, and in identifying the players, it was clear from the outset that the 'developer' and would-be employer of contractors and professionals might on analysis be a poor substitute for the cash-rich household name, albeit attended by bankers, institutions and tenants, the last-mentioned of whose prospective income commitment, secured by a developer/tenant agreement for lease underpins the whole financial edifice. What we probably have is a shell company, a single-purpose vehicle, to satisfy banking requirements, out of whose, probably small, authorised capital only a handful of shares are actually issued. Not that this book will go one inch of the way to achieving any alteration of the balance between the power of the employer on the one hand and of his contractor and professionals, on the other. Commercial considerations override.

That lawyers have, particularly in recent years, burrowed their way into the process is no accident, and the more curious reader should turn to a specialist work. The upshot, however, is that forms of building contract, appointments, warranties, guarantees and bonds have all been influenced, first, by erosion of their predecessor forms' content through amendment, leading to shortfall in legal remedy and, second, by the recognition of the existence of a number of third party concerns and rights which cannot necessarily be satisfied in law by a remedy against, say, a principal developer who is also an employer of construction. The Insolvency Act of 1986 marked a turning point in providing further employer protection of a certain kind, by channelling employers into the use of SPV companies, albeit now further channelled by the Enterprise Act 2002, as mentioned earlier, and with the Construction Act, as it is shortly known, otherwise the Housing Grants, Construction and Regeneration Act of 1996, a further turning point still, perhaps representing the first, albeit wholly unrelated, step in the opposite direction.

Other events in recent years such as the Construction (Design and Management) Regulations 1994 (as now amended), the construction industry scheme (replacing the earlier statutory tax deduction scheme) as indeed provided for in the joint contracts tribunal (JCT) and other standard forms of building contract and the scheme for construction contracts (giving effect to Construction Act provisions where found wanting in construction contracts) have all had their impact. Throw in for good measure the Contracts

(Rights of Third Parties) Act 1999 and what we say, what we do and how we do it is already very different from what it was even a few years ago. Even basic forms of contracts are changing, and the JCT major works form, introduced in 2003 is an intelligent advance with enormous potential, in the right hands. Time will come, one imagines, when with thought and diligence it will lend itself to smaller projects and influence the redesign of other forms.

However, whilst the construction industry and professional bodies have, with the aid of learned advice, embraced changes in the underlying law with fervour, development and construction lawyers must have regard to the agendas of those who are likely to be influenced by or benefit from the construction process, which is what essentially drives amendments to standard forms of building contract, prescribed forms of appointment, warranty and so on. However cleverly contrived, with notions of fairness and caution in mind, the carefully crafted forms produced by the contracting and professional arms of the industry, with their particular protection in mind, will usually yield to consumer demand, particularly the institutional market. In other words, if you want to be paid, follow the money.

Standard forms at large, having been so carefully contrived by the industry bodies concerned can thus, for that very reason, provoke the institutional market. As between employers and their employed, properly advised employers look to consistency of obligation on the parts of those employed, limitation of employers' obligations, removal on limitations of the employed's obligations and, last but not least, linkage of disputes. It is helpful, however, before crying foul, to remember who is paying whom. The result in the case of professional appointments will tend to be the abandonment of standard printed forms, retention at best of identifiable professional duties, and the setting of a different agenda consistent across all related appointments. Moreover, albeit that some employers also use their own forms of building contract, standard forms such as the JCT family will also be heavily amended in line.

This chapter addresses some of those issues and may even be of help to insurers in the course of the day-to-day tedium of approving documentation which their insured clients may feel disinclined to submit to legal advice, often for fear of swallowing up the fees which they expect to earn from their appointments. The rationale for amendment or the use of a new form also is so often not understood by those lacking legal qualifications and the requisite experience, to the point where time and again amendments are even sought by them to greater disadvantage. Lack of comprehension is a dangerous thing and, unhappily, the fault line sometimes lies at the level of basic literacy, and has absolutely nothing at all to do with fine points of law.

Much can be achieved by careful reading and analysis of the subject document, and by not dismissing a proposition as so much jargon. The reality is that courts dislike jargon too and are looking for meaning for, rest assured, the results of a dispute or default will be very meaningful indeed. Idleness in composing and expressing terms of engagement is simple folly. Nonetheless, so often negotiation with non-lawyers, and occasionally it must be said with lawyers too, descends to bizarre levels, in the sure knowledge sometimes that the person one is dealing with has no prospect of understanding the proposition in hand. Persistent failure to comprehend suggests that at the very least those who are leading the appointment/contract should not delegate the small print. The deal has not yet crystallised.

It is thus a job for the senior people who struck the deal in the first place. It is helpful to remember at all times that what has been written may one day be subjected to withering judicial scrutiny. If in doubt, take (good) legal advice.

## The institutional mind

The single developer or owner, financially supported from own resources, and perhaps developing for own use or speculatively, may at one level have only his own interests at heart or, perhaps, he will need to consider institutional requirements in terms of characteristics that others may look for, if not now, then at some time in the future. If a development is brought together on the back of one or more development agreements it has already been seen that a development agreement must contemplate construction but, in turn, unless the construction documents meet institutional requirements, then they too will be rejected. Few construction professionals have the time or the inclination to see what drives the development process or to examine the rationale for requirements being laid down in development agreements, and innocence abounds. Hopefully, this chapter will make it a little easier.

Starting with the classic landowner/developer agreement, precisely what is it, therefore, which the landowner is seeking in prescribing his development requirements? It is a first lesson for would-be development lawyers to understand that development agreements should not be regarded as all the same and that, in every case, the objectives and performance criteria should be led by the single criterion of the landowner's interest in the built outcome. Unless achieved, the developer's ambitions for himself will never be realised, which is perhaps a sobering thought.

The developer's solicitor, negotiating that agreement, should use the criteria so identified as his yardstick for negotiating out and eradicating those requirements which the landowner has incorporated but which do not, in substance, touch or concern his interest. Let us assume, however, that the interest in question is substantial and perhaps complex. This author has in mind, as an example, a transaction some years ago for the body then known as London Regional Transport concerning a major development comprising in part and otherwise intertwined with a transport interchange. Land would be made available for major commercial development including offices, a shopping centre and below-ground car parking: the consideration for the land was expressed partially in cash including overage based on disposals, and in part deemed disposals, but otherwise for the landowner's particular benefit in terms of providing a revitalised underground station and a new bus station as part of an interlinked development. The major interest to be taken by the developer was, therefore, inevitably a long lease, not merely to facilitate and manage the physical division of interests both horizontally and vertically, but also to facilitate the operation of a regime for major servicing, such as structural repair. Given that part of the consideration was reflected in a complex overage formula based on disposals, and in particular that the transport interchange was to be provided by the developer at its own expense, it is not difficult to see that the landowner's interests were substantial indeed, including in its capacity as a transport authority. Whilst transport led/dependent development has additional characteristics of its own, the commercial interests inevitably dictate the necessity for certain requirements. In the nature of such development, similar requirements are also to be found in town centre schemes and other complex developments.

The primary *sine qua non* is one of design and purpose: thus, always with timing in mind, one finds design criteria laid down but, before one gets that far, there are first of all even more basic considerations. For a professional to be appointed at all, the expectation is that criteria will be prescribed within a development agreement both as to identity of the professional team and contractors and as to the terms of engagement – the latter are dealt with further below. This implies a tendering process and the prospect that within that the landowner may require that certain professionals with whom it has worked happily before be invited to tender and, possibly, others may be excluded. One does not expect to see an express obligation on the developer that thou shalt not employ so and so, even if it is said privately in meetings, but one does expect to see a requirement, perhaps, that the developer shall shortlist so many out of whom, with the consent of the landowner (usually not to be unreasonably withheld), so many are to be invited to tender, and the form of appointment will also require approval. Expect funders to do the same, so that the detailed terms fit their own agendas. Some landowners, on the back of past experience, will simply not feel comfortable with this or that firm, and the developer will simply have to go along with it. Naturally, where complex development is concerned, by the time matters have reached the point where solicitors are to be instructed to document the transaction, so much of the general ground work for the project will have been discussed and agreed upon that there will be few surprises. The development agreement sets out the parameters for fulfilment of each side's objectives.

So far as concerns contractors, there is perhaps an even greater need for understanding at the outset, before prescribing how they are to be appointed and on what terms. However, one should expect that the developer will be required to produce a shortlist, perhaps including one or more nominees of the landowner as a basis on which to seek tenders.

The appointment of professionals, particularly quantity surveyors, architects and engineers is of course vital to the early stages of the development, the preparation of plans and the seeking of planning permission and other approvals, not to speak of costings which underpin (or undermine) the players' financial objectives. Preliminary appointments may also be required before those for the main project. The plans themselves and the terms of planning permission, for example, go the root of the landowner's objectives and will again be expected to be subject to an approvals process.

None of these things simply 'happen', they are all the subject of contractual obligation and what the professional team do is, therefore, subservient to it. Worse, the team may not be exposed to any part of this process, relying solely on what they are told they must do by the employer. Accordingly, in the fullness of time, a project emerges which meets the development objectives of both landowner and developer, funders and other key players.

Once it is a buildable scheme, the next stage is implementation. Enter the chosen ones (professional team and contractor). Again the terms of engagement and of the building contract are subservient to contractually expressed objectives. They may be subservient on the basis that what has been contractually prescribed coincidentally meets the requirements of the development agreement, or the terms of the contract are expressed to be subservient to it. That is why, for example, the texts of one or more development agreements are so often to be found comprised in building contract requirements, and careful drafting is required to ensure that there is clarity in interpretation where a conflict of intention could emerge.

It is, in fact, the product of lazy draftsmanship as so many development agreements are written without a great knowledge of construction, and the potential for conflict is great. If the terms of a development agreement are to be incorporated by reference, be sure to have the contract identify the specific provisions referred to, and take advice on the impact.

Whether or not incorporated by reference, development agreements also prescribe performance criteria: these are broadly divided into the requirements of the development itself and then the process of meeting those requirements, say through supervision and certification. A landowner's representative, appointed by the landowner under the development agreement is there to measure the developer's performance. Whilst the developer is the employer of the contractor and the professional team, the landowner's surveyor thus has no direct supervisory role, the developer's solicitor (one hopes) making sure in particular to provide that the surveyor, whilst entitled to enquire about and view progress, has expressly no right to issue instructions. Nor, if the agreement is properly drawn, should the surveyor assume any other part of the developer's function as employer, e.g. by operating as his agent, save expressly, for example, during a period of unremedied default on the part of the developer after formal step-in. Usually, however, there is no halfway house and, whilst the developer is the employer, that is the end of the matter unless and until there is an outright break, and step-in arrangements, as contractually prescribed, are brought to bear. Therein, accordingly, lies some of the rationale for assignment provisions in appointments and step-in provisions in warranties, about which again see further below. However, as the particular kind of development agreement also caters for limited alienation under controlled conditions, step-in (if at all) by the landowner is an ultimate if remote remedy. The important thing to understand is that step-in must notwithstanding be prescribed, even though bank lenders will first have been given the opportunity to protect their own position, and this means alienation under controlled conditions. No wonder then that alienation provisions in building contracts are re-written to accord.

At all events, by virtue of the development agreement, the developer will thus have been steered in the direction of creating his construction relationships in a certain way, and when lenders, investment buyers and tenants become involved, the need for careful interaction between those contractual relationships is even more marked, for fear that the ability to reach satisfactory agreements with those involved in construction may be impaired or made simply impossible. Where a disposal is in mind, say an agreement for lease to a tenant of the completed development, there is, of course, a better way for development lawyers to proceed: devise a construction regime and build the development agreement around that, instead of the other way round. Regrettably, some are only just learning.

## Banking considerations

If the development is led by bank finance, a further raft of development objectives and performance criteria will be found in the enabling loan agreement. First, the landowner/developer agreement, if there is one, will have had to be compiled with financial requirements in mind. Second, all of the matters discussed above may, to a greater or lesser extent, require a similar approvals process, followed by implementation and satisfaction of performance criteria. This represents a substantial burden for the developer. Suppose, for example, that an interim certificate has been issued by the contract administrator under the building contract; in other words, he is satisfied that a certain stage has been reached in accordance with the terms of the building contract. Suppose, then, that, under the terms of whichever development agreement, it is the view of the landowner's representative or the lender's representative that the particular stage has not been reached, in accordance with the terms of that particular development agreement.

First, all the contracts, including the building contract, should have been drawn in such a way as not to invite such a conflict. Second, the independence of the contract administrator (albeit vulnerable to attack via the employer), must not be compromised. The developer cannot be heard to require that his landowner is not satisfied and, therefore say, on those grounds the contract administrator must not issue his certificate. If the certificate was properly issued, so far as it went, but performance criteria in the relevant development agreement had not actually been met, the inference is that the developer must take further steps, at his own expense, to meet those criteria. It is a *sine qua non*, therefore, that the developer should not, in his negotiation of development agreements, invite that sort of conflict. That he may do so is a problem of his own making (or of his lawyers), and there it usually lies.

It is thus better for development obligations to be drawn when possible to equate with what, after the tendering and approvals process, the developer should prescribe in his contractual relationships with those involved in the construction process. After all, the detailed terms will have had to be approved, and if acceptable in construction terms, then why not adopt them as the standard to the development. Ultimately, either the developer is in breach or he is not, and if he is the question arises as to whether or not he should initiate a construction dispute with those whose work is at fault. This, of course, may involve the contract administrator as a party. It should be remembered at all times, of course, that whilst the landowner has an interest in the built outcome, the lender has an interest principally in being repaid, and that means, inevitably, disposal of the problem and the introduction of new players, unless the bank can the more readily (and profitably) see the project through.

The institutional investor, in turn, has an (eventual, if not immediate) interest as a landowner concerned with the built outcome in terms of his actually owning the resultant investment, such as under a forward funding agreement. If his agreement with the developer is forward sale/purchase, his remedy is one of walkaway until he completes his purchase. The construction team may not feel his presence, for the most part, unless and until the project is completed and warranties are required to be given, if not an outright assignment of the building contract and appointments unless the burden of responsibility is to be left with the developer. By contrast, the investor under forward funding is a landowner from the outset and remains so to the exclusion of grant of any major interest in favour of a developer. The developer is essentially a contractor (being also treated so for tax purposes) and the investor is thus, very much, the owner of the development.

It is therefore time to put away the niceties of construction law, and all attempts to channel the client into one's perception of what is or should be a balanced approach, and face the realities of this life. We call it the market place.

## Meeting institutional requirements

### Building contract

Those used to dealing with lawyer-prepared documents will notice a general pattern of contract amendments. These are driven in part by individual requirements but more so by evolving 'standard' amendments, themselves often driven by precedents offered by legal textbook writers. They are an invaluable aid to employers but, as always, require intelligent use.

Where a contract is entered into within the jurisdiction, concerning a domestic subject matter, it is hardly necessary to state that the proper law is

English law or that the courts of England shall have jurisdiction: they will anyway. Because of legal limitation periods (of duration of enforceability), expect most contracts and appointments, compliant with related development agreements, to be executed rather than signed under hand. Institutions properly advised will have nothing less.

When reading a building contract always read the (revised) definitions section and where amendments have been made, take care to read and assess them. They are fundamental to interpretation. Incidentally, the JCT major works form offers a definition of practical completion, a concept which has hitherto been essentially a matter of judgement for the contract administrator, although there were available textbook definitions before. Not that he can thus cease to exercise judgement, but his judgement must instead be directed towards satisfaction of the definition. Once the definition has been complied with, in his judgement, he cannot go beyond the definition and apply some standard of his own, except insofar as the contract actually allows him so to do. In practice, definitions of practical completion are increasingly used, and the adoption at last by the JCT in the major works form follows the trend. The JCT offering also leaves some flexibility for the contract administrator by prescribing that 'Practical Completion takes place when the Project is complete for all practical purposes *and in particular* ...'.

If the contract is, say, in the design and build form, or there is a design supplement, then employers will be looking for something similar to design warranties in professional appointments discussed below. Basically, one expects to see a warranty within the body of the contract that the contractor has used and will use all skill and care (reasonably) to be expected of a duly qualified and experienced designer. Use of the word 'reasonably' in context may be a little over-subtle, because a fairly objective standard is set by the very fact of professional qualification and experience. The JCT major form now follows suit (but without inserting 'reasonable' anywhere), adding that the contractor does not warrant that the project when constructed in accordance with its designs will be suitable for any particular purpose. This, of course, begs questions, as fitness for purpose warranties are often sought. They may be the product of a lawyer's ego (let us be honest) or they may be directed towards a real need. It is important to identify which. However, once notions of fitness for purpose are introduced, and unless tempered by express exclusions, the shoals and banks of consequential loss come ever closer. Beware.

Standards are also applied to materials used, with reference to the employer's requirements, say, but contract amendments may prescribe that they are reasonably suitable for their purposes. The JCT major projects form adopts the notions of its predecessors but adds 'or, if no such kinds or standards are described, materials and goods that are reasonably fit for their intended purpose' adding that all materials and goods shall be of 'satisfactory quality'. Similar sentiments are applied to workmanship.

Practical completion clauses invite amendment, and this may be driven by tensions caused by development agreements, with interested parties jostling each other for evidence that performance criteria have been met. The same applies to an extent to interim certificates and demands for payment, and underpins the need for the developer in his other dealings to peg performance criteria so far as possible solely to the conduct of the building contract. It is of more than passing interest to the contractor, after all it may mean that in consequence there may be nothing forthcoming with which to pay him, if the developer is drawing down funds under a different set of rules. Unfortunately this is a classic case of development and construction being poles apart. It need not be so, but has little prospect of improvement while the two are kept apart.

One danger for the developer, and in particular his joint venture and funding partners, is premature certification. It is potentially deterred by defining practical completion, but meeting performance criteria in development agreements constitutes much of the driving force behind forms of appointment and warranty drawn to satisfy the institutional market. It is fair to say, however, that, since the 1990s' recession, this has become less of an issue.

Quite apart from the changes in the law which have altered the face of construction documents so much in recent years, the three-line letter referring vaguely, say, to RIBA terms and conditions, and intended to constitute an architect's 'appointment' disappeared from his files long ago, and indeed for the most part before the 1990s' recession. At least one hopes that it did, and so should the architect. Similarly the developer who calls up his friendly architect to bring forward practical completion to hasten that all important draw-down to enable him to meet a variety of commitments, will no longer impress his bankers who, for a long time now, have armed themselves with an impressive array of loan agreement conditions and direct warranties from the professional team. Legitimate snagging lists cannot be avoided, nor usually, during the defects liability period, the raising of lists of defects. These will not be left to the contract administrator alone although he will not be denied his judgement in the matter, and parties to development agreements will monitor carefully and present the developer with their requirements, which are duly collated. There may also be arguments behind the scenes kept hidden from the construction team, but the practicality is that the other parties' representatives will also be working closely with the professional team and the contractor, and it is essential that they all work well together.

The coordinating role of a project manager is thus often pivotal to the success of a project, both in foreseeing difficulty and in handling it tactfully when it arises. Even a lawyer understands that if you can manage the people and personalities, the less likely you will have to look to the small print. The danger lies in relying on goodwill and commonsense as a substitute. Do not!

Partial possession and access provisions are frequent candidates for contract amendment as needs often vary. Phased development apart, and even if within the terms of the building contract the contractor is able, in practical terms, to acquiesce in partial handover (thus signifying in respect of the part in question that it is actually practically complete) take-up of occupation may be impractical. The implication is that the tenant's own team will want to start fitting out, they will need facilities perhaps by no means confined to the unit of accommodation and they will be perhaps be under the feet of the contractor and the sub-contractors wherever they move. Depending on the scope of occupier's works there may be overlapping insurance issues too. Expect to find amendments to standard form contracts modifying provisions for early employer access. Likely as not, where there is a developer/tenant agreement in play, that document will contain a raft of provisions detailing the relationship of developer and tenant in relation to those works.

Indeed, the development provisions of an agreement for lease between developer and tenant will likely cater, in some considerable detail, for the operations necessary to enable the tenant to open and trade as soon as possible. Everything is driven by timing and practical measures have to be taken to accommodate tenants' requirements. In a multi-occupation development, it is not just a matter of connection of the utilities but of facilitating the to-ing and fro-ing of workmen with their vehicles, and storing of materials if not within the unit then close by. There may be much work which can be done before practical completion as such is certified, thus implying that partial possession may fall far short of an answer. If the development is based on one or more pre-lets the opportunity arises to accommodate all of an individual tenant's requirements, as requested in his enabling agreement for lease, into the employer's requirements. Where the agreement for lease follows the building contract instead, then the building contract should contain amendments imagining so far as possible what a tenant's requirements are likely to be. The developer who fails to cater for this is only storing up problems for himself. Contractors who will not countenance such 'interference' may only be edging themselves out of a job. Practicalities are key.

The next common area of amendment lies in the alienation provisions, that is to say assignment and sub-contracts. For institutional purposes, provisions such as at 18.1 of the JCT design and build (i.e. JCT with contractor's design) that 'neither the Employer nor the Contactor shall, without the consent of the other, assign this Contract' will not do. There is no implication of law in a building contract (unlike a lease, say) to imply in any way that such consent is not to be unreasonably withheld. Consent is thus voluntary unless expressed. The funder or landowner, by contrast, standing behind the developer, needs to be able to deal as part of the fabric of securing its interests. Expect no freedom for the contractor, but expect a requirement for complete freedom for the employer. Resist too far, and it simply renders the scheme 'unbankable'. Perhaps not surprisingly, the JCT major works form now encapsulates these requirements precisely.

What the major works form does not cater for in detail is insurance, the parties being left to be grown up and prescribe their insurance needs for themselves. In Chapter 9 we look at insurance and risk, as required not just for construction but for the development operation as a whole. After all, the risk lies not just in the physical project, everything may run perfectly until a delay causes delay in turn to the lead tenant in putting back his commencement of trade, thus also deferring income to the investor, and so on.

Where insurance is dealt with in the standard forms, expect at the very least for interested parties' names to require to be added to the contractor's insurance as joint insured and that, given the kinds of interests that they have under a variety of development agreements, they will not be satisfied with the response often given that this cannot be done because of the contractor's block policy arrangements. In the case of institutional development, it will be done, because they say so (but note the impact of this as explained in Chapter 9). Incidentally, the extent of interim access by tenants may come under the microscope in the context of insurance, and even where partial possession is in contemplation to enable occupation and fitting out, the matter should be explored fully with insurers to ensure that the parameters of exclusion are not breached and that if the contract does not so provide, then arrangements are put in place by interested parties until mainstream buildings insurance is in place.

Delay and extensions of time are an obvious target for institutional concerns. Contractors' excuses, sometimes more politely termed relevant events, will come under scrutiny and will likely be the subject of substantial deletions, underlining (to avoid defining the implicit) that no extension of time can be allowed when the need for it was the contractor's fault in the first place. A sobering thought, as the guidance note to the JCT major works form observes, the 'list is somewhat narrower than is found in other JCT forms and excludes exceptionally adverse weather conditions, industrial disputes, the inability to obtain labour and/or materials and delays caused by statutory authorities'. That's the spirit! It is also a familiar pattern to amendments to the other JCT forms. In the event of default or some event giving rise to an employer's right to determine, remember that he in turn is led by the provisions of relevant development agreements. First, he will not usually be free to make that decision except with the acquiescence of other principals. Moreover, neither will he be free to appoint another contractor without going through the same procedures as applied to him, under the development agreement, for the initial appointments. The whole process of termination and a new appointment is thus heavily choreographed, even though the contractor and professionals may be oblivious to it. Similarly, as will be seen below in connection with warranties, whatever the provisions of the contract, neither may the contractor be entirely free to terminate unless he first complies with his warranty obligations. These are meaningful documents. It may be that he has simply not been paid: funding and loan agreements habitually contain provisions facilitating direct payment for that very reason. However, this is not always so, and major developers may have negotiated direct draw-downs to the SPV, often precisely to exercise some leverage. From a funder's perspective, there is a clear danger to the project, and judgement must be exercised accordingly.

On the subject of payment generally, retentions have long been a source of contract amendment (conveniently the JCT major works form does not provide for them and, if required, they need to be added back). Otherwise, to protect the contractor the retention is required to be held on a fiduciary basis, i.e. the employer effectively becomes the trustee, according to the printed forms. Whilst the banker may respect that notion, allowing it to be drawn down and placed in a designated account, the likelihood is that it will be retained by the banker and included in the draw-down at the appropriate time. Apart from anything else, this also suits the developer, as he thereby avoids paying interest while the retention sums remain undrawn. The developer is thus prospectively in breach, and contractors would be wise to flag the point up and ask for evidence of compliance, particularly when the funding sources are not known. The same sentiments apply in the case of an institution under a forward funding agreement.

Accordingly, expect a contract amendment to provide that the employer will not be under any fiduciary obligation concerning the retention and, in particular, that he will not be obliged to set it aside in a separate account. From the contractor's perspective, he may perhaps take some comfort from the fact that, if it hasn't been drawn down it remains safe with the bank or financial institution in question, which is more than likely going to play fair when the amount becomes due.

Standard provisions concerning antiquities do not usually invite amendment. If the employer is not himself the landowner, the inference is that there will be provisions somewhere in a development agreement to which the employer is subject and which will govern any instructions he gives to the contractor. The developer, if appropriately advised, will have made all necessary enquiries but major finds are inevitable from time to time and will be subjected to investigation and consequential delay.

Expect changes to adjudication clauses: nothing must fall outside Construction Act provisions of course. Whilst the JCT approach is that the adjudicator shall not be obliged to give reasons with his decision, a likely amendment will be that he shall. It is helpful to expressly empower the adjudicator to determine more than one dispute at the same time (to be distinguished from a power to join other parties to a dispute) and, indeed, to determine any matter raised by a respondent party by way of set-off, abatement or counter-claim. Such provisions will likely be found to be interwoven with professional appointments. An obligation to determine other matters carries with it the possibility that the other matter may, in the event, not be the proper subject of a construction contract, albeit that for practical purposes such a possibility may be fairly remote.

Finally, whilst retreating a little from the merits of arbitration versus legal proceedings it is helpful to remember that when documents are silent on disputes, or silent beyond adjudication, then the natural resort thereafter is legal proceedings. If the arbitration route to ultimate dispute resolution is chosen and according to whatever industry/professional body rules are used, subject to the Arbitration Act 1996, joinder of related proceedings needs to be advanced from power of the arbitrator to the express contractual rights of the parties in each case to require it, throughout. Lawyers often favour final determination by the court alone, but that is a better subject for a book on construction law, not development.

## Professional appointments

By now, it is no longer necessary to underline that the institutional market works in a very different way from the domestic relationship arising between client and professional or contractor, still less the way the construction part of the equation would like it to be. A glance at, for example, the conditions of engagement under standard architects and engineers forms, simply underlines the problem. A series of obligations points to an intended measure of protection for the professional. That is not to say that the institutional approach removes all protections, so much as leaving it to the professional to prove the implied condition that arises, say, if his instructions are untimely or incomplete or that the client has not engaged others, or conducted those engagements, in a way which facilitates the professional's functions. Standard forms, for example, inhibit assignment by the employer. The institutional and banking attitude is that as with standard forms of building contract it is none of the professional's business. Here is a project with a complex financial structure and no dealing with the business objectives of the principals concerned must be inhibited or prohibited by anyone engaged on the project.

Greater sympathy will lie, however, in the defining of services, in the nature of which the employer relies for the expertise of the professional, but this is not to say that services will not come under the scrutiny of the employer, often aided by his quantity surveyor as well as his solicitor in prescribing them for the job in hand.

Containment of liability is another area of difficulty. Attempts to limit the liability to the extent of the available professional indemnity insurance do not always find favour any more than the antithesis of a fishing expedition implied by what are known as net contribution clauses which set up assumptions that, say, the contractor and other professionals have been engaged on no less onerous terms. Not that such provisions on their own avoid or reduce liability: negligence can be difficult to prove and the fact that someone else happened to have a related and similar obligation in relation to his own function cannot, at a stroke, relieve the burden. As always, it depends on the circumstances. Accordingly, whilst pure construction lawyers instructed by contractors and professionals will seek to bind in a number of safeguards, the agenda of the principals standing behind the employer is rather different. Professional negligence causing economic loss, and breach of contract, causing consequential loss, may be tightly intertwined and it has to be said that under most development agreements the developer himself will be afforded no comfort at all by his own master. It is hardly surprising therefore that in his dealings with the contractor and professionals, what is prescribed in their terms of engagement may simply reflect that position.

Expect, therefore, when dealing with development of institutional quality, to be presented with a more formal arrangement, including any services performed prior to the agreement being entered into being brought into account as having been performed under the terms of the agreement: a back-handed approach to extraction of an indemnity arising out of imperfect prior performance. So many of a consultant's obligations will have been performed in preliminary stages, not least just to get planning permission, at a time when his 'appointment' may amount to no more than a design brief, and so those functions will be covered as well as those which would otherwise be subject to the terms of the appointment, but for the time it took to get signatures on paper.

Use of skill and care provisions, much as described above in relation to designed portions of a building contract will also appear, no doubt expressly excluding the developer from responsibility for omissions on his part to approve or disapprove designs and so on. Exclusion clauses are, however, likely not to work, as a matter of law, against negligence, although that will cut no ice in negotiation if applied by the professional to his own obligations. Do not expect to achieve much in making protective amendments of your own of that nature (but try if you can).

Both construction briefs, comprising employer's requirements or whatever, and conditions/appointments may thus draw attention to related development agreements (see earlier chapters), containing provisions intended to ensure that the contractor or consultant does not by act or omission place the developer in breach. They may also go on to prescribe additional certification and, perhaps more controversially, compliance with procedures such as consultation with a bank's representative before draw-down, prior to issuing a related certificate. It is necessary to make a distinction between professional obligation or discretion and apparent subjection to the influence of third parties. This is particularly important in the case of the contract administrator who must operate with a measure of impartiality and be seen to act fairly. Unless (unlikely) the building contract itself says a certificate shall not be issued until some procedure required by a third party has been complied with, then the contract administrator must issue the certificate expected of him. Accordingly, what is actually being asked of him is not to delay or pervert the course of exercise of his judgement, but to ensure that he has first done what is expected of him. If such provisions appear, say in an architect's appointment, it may be helpful to express such obligations as being expressly without prejudice to his obligations as contract administrator under the building contract, in so many words. They may be the subject of careless shorthand and are best handled with the benefit of legal advice.

Design changes can cause difficulty depending on the extent of authority afforded to the professional. Even if they seem desirable, the institutional hand on the consultant's warranty if not the initial appointment will likely reflect that the professional has been forbidden any flexibility which would affect lettable area or investment value.

The professional team will inevitably be required to work with each other's designs, including those of specialist sub-contractors. The architect, say, may be asked to review these and integration at large sometimes provokes fears of responsibility for others' work. Vicarious liability should be expressly excluded so that only originators of designs are responsible for them. There is a limit to what can be expected but, within that limit, even then the prospect of negligence, as ever ultimately inescapable, can arise. There is often some sensibility also over the extent to which designs may be used for other purposes connected with the development: it is an over-simplification, but those sensibilities are sometimes overstated, as no-one is usually expecting a professional to anticipate the unknown, and what cannot be contemplated is a candidate for avoidance of legal remedy.

Throughout, the developer is looking for a consistency of approach in all his appointments to link with related development agreements and to enable him to deal effectively with the investment in due course, the other parties to those agreements also. Consistency of obligation also stretches as far as consistency of insurance obligations, requiring those with a professional responsibility, including contractors with a designed portion, to carry personal indemnity (PI) cover at appropriate levels for a period usually of up to 12 years from practical completion (or from termination of the appointment if earlier). There is, of course, only so much burden that can be borne and commercially reasonable terms, premium rates and so on are fundamental and, where reasonable bounds are breached, then expect a procedure to be prescribed for whatever is available and would be fair and reasonable in the circumstances, or negotiate one.

And now a further word about assignment. As has been seen, none of the professional team or indeed the contractor should be placed in the privileged position of interfering with the business relationship upon which, indeed, his appointment relies. As with the building contract, expect freedom to assign to a mortgagee without consent at all. Assignment to a mortgagee is part of the security package from which there must also be an exit route to facilitate repayment of the loan. Both the funder's exit and assignment by the employer in any event are similar areas of difficulty. Some PI cover and insurers' representatives themselves can be unhelpful, granted, but unduly restrictive cover may possibly impact on employability.

There can be other areas of difficulty for the professional too, when suddenly faced with a developer (or his SPV) with whom he fell out last time and never wants to work with again. Some draftsmen cater for this on the basis of allowing reasonable objection to assignment, not with a view to empowering the professional but with the view to termination of the engagement so that a new professional can be employed. An express ability to pay off the professional at any time is the employer's answer but, as with everything else, the professional must take his own view. He must take it or leave it. Subletting of services by the professional is usually also expressly excluded.

Issues of copyright have been discussed above in relation to building contracts, especially in relation to whether or not designs can be used for a variety of purposes including extensions. A royalty-free licence will likely be granted which is capable of subletting, supported by provisions such as a warranty that the use of the material will not infringe rights of others.

Construction Act provisions and consistency of disputes have been otherwise aired above, but before leaving appointments, it is as well to visit briefly third party rights which may arise under the Contracts (Rights of Third Parties) Act 1999. The practice until now has been, indeed in standard form contracts and terms of engagement, to exclude the effect of the act entirely so that only the original parties or permitted assigns can benefit. A change of attitude could only have come sooner or later, as indeed reflected in the JCT major works form which, at clause 30, introduces funders' rights and purchasers' and tenants' rights, purchasers and tenants nonetheless having expressly no authority to issue any directions or instructions to the contractor. It is, accordingly, only a small step to the re-working of third party rights clauses to express a building contract or professional appointment in terms that it will benefit express classes of third party, perhaps in ways closely analogous to those which would otherwise be found in a separate form of warranty. The facility has existed since the act came into force but, as with everything else, the idea takes time to develop, and perhaps unfounded fears need to be calmed.

### Warranties

If advances are made so rapidly in embracing the changes implied by the 1999 act, this section of the book may soon be redundant. However, for the moment, construction warranties still feature as an integral part of any construction package. They allow remedies to a variety of interested parties, in some cases facilitating substitution of the beneficiary as employer under what is, in effect, a form of novation. The extent and nature of a person's interest may be self-evident, sometimes it is not because of a unique contractual relationship with the developer. However, just as the disposal or (say, in the case of a lease) alienation of an interest leaves the beneficiary of a warranty with no enforceable interest in that warranty, neither is it of any value to someone who has no interest, in the legal sense, at the outset. No damage, no damages.

A warranty will, therefore, usually recite the interest in question upon which it is based and, usually also, will immediately launch in its first operative provision into the duty of care which the warranty is intended to underpin. It will prescribe for something like all the professional skill and care and diligence reasonably to be expected of a suitably qualified and experienced consultant in performing services such as those prescribed by the appointment. This sometimes provokes a string of amendments, born of fear that somehow the professional's liabilities have suddenly been increased, for the most part unlikely unless that would be the clear consequence. Not so, at least to the extent of the example above. It is also helpful to remember that damage is true loss and damages follow accordingly. English contract law makes penalties void, and, as an aside, the error of imposition of a penalty in any contract is likely to be avoided by reference instead to liquidated damages, as being what the parties have actually agreed is the prospective loss. Lest there may yet be doubt, an amendment allowing the professional the same rights in defence of proceedings as he would have under the original appointment, and so on, are worth adding for comfort, and fair employers and potential warrantees will allow it.

Insurance provisions will likely follow those of the appointment including for the provision of documentary evidence to the beneficiary of the warranty. Warranties otherwise usually contain independent insurance clauses, but this is more for convenience and visibility as anything else. If the warranty says the professional will perform the obligations in his appointment, it necessarily includes his obligation within to insure, unless it is expressly excluded.

A warranty essentially leads to a remedy in damages: it is not a substitute for assignment of the appointment of contract in question. A contract is an entire entity, and one cannot assign the same subject matter twice (although, but for express prohibition, assignment in parts is another matter and fragmentation can be problematical. A fair employer will not usually mind being restricted in his own assignments to assignment of the whole, on the basis that it can rarely be contemplated otherwise). Whoever has the benefit of the contract, himself perhaps an assignee, additional beneficiaries can thus only have a warranty. Thus, the employer (developer) employs, and, ideas of step-in apart, the underlying landowner, the funder, or a purchaser under a forward sale agreement, or a tenant under an agreement for lease, all pick up warranties, the last two mentioned usually when their major interest materialises in the form of purchase of the investment and completion of the lease in question. The developer remains as employer, probably because he is required to do so as part of his contractual relationship with other principals, to deal with residual construction issues, perhaps during a time when he is still contractually obliged to seek tenants. Eventually, he may be required to assign, say, to the purchaser.

However, as mentioned above, within forms of warranty may be found step-in arrangements, amounting in substance to a novation. Those entitled to step-in will usually prescribe that the contractor/professional will not terminate without notifying him, nor discontinue services, say because the developer is in some breach, and the warranty then triggers a procedure for step-in/novation, usually accompanied by making up any shortfall in fees as a precursor. This allows the landowner, perhaps an institutional purchaser under a forward funding agreement, to resume the development upon failure of his developer. It also allows a mortgagee, in case of developer breach, to put in an administrative receiver, or administrator as the case may be since the Enterprise Act 2002. Where there is an underlying landowner/developer agreement, the default provisions mean inevitably that the mortgagee must dispose and bring in a new developer under controlled conditions. It is often helpful to modify step-ins so that a conflict does not arise in relation to other parties. So a bank might require a developer to restrict the ability of a tenant of a single tenant development (substituting himself as developer) to implement substitution without written approval.

With many developments, the construction documents are themselves part of the security and so a contract or appointment may be assigned to the lender, without need for a warranty, subject to the developer's right of reassignment upon repayment of the loan. If the developer is in default, the comprehensive security taken by the lender over every aspect of the project means that he can thus deal on to a third party, as if he were standing in the shoes of the developer (his insolvency representative if he continues the development being expressly appointed as attorney of the developer for the purpose). The developer must in turn ensure contractually that he is permitted to operate the appointment until, say, an event of default on his part. Reference will sometimes be found in banking documents to 'security assignments'. Because a charge over contracts cannot work like a charge over land, which depends on statutory requirements for modern type security, security over a contract is essentially by assignment coupled with the right of redemption. Once assigned, however, the assignor cannot usually enforce it without the kind of device mentioned earlier. Other banking requirements prescribe only for an equitable charge over contracts and appointments, under which the employer's rights of management and enforcement continue notwithstanding.

The rationale for limiting assignments at all is not wholly convincing: if, say, a lease is assigned several times during the period of the warranty (say 12 years from practical completion where the warranty is executed as a deed), but if perhaps only two assignments are permitted, it means that the benefit of the warranty may expire sooner rather than later upon the third assignment of the lease. Indeed, this defies a legal rationale but limitation of assignments seems generally accepted.

The inclusion of a copyright clause essentially confers the benefit on the beneficiary of a licence to use the material, so it is not sufficient to rely solely on the appointment which, in turn, conferred a licence on the employer. An operative provision is accordingly required.

Finally, some industry forms of warranty include the very clauses limiting liability which are mentioned above, including in relation to net contribution. The origin of net contribution clauses is the Civil Liability (Contribution) Act 1978 under which someone who is liable for a loss may recover a just and equitable contribution from another person who is liable in respect of the same damage. This does not have effect to reduce his liability but does allow for the claim of a contribution from a third party. Net contribution clauses purport to set up an assumption, whether or not the case, in line with the act so as to limit the liability. Given that the Unfair Contract Terms Act 1977 provides that someone cannot restrict liability by reference to a contract term save insofar as that term is reasonable, the prospects are at least interesting. When this chapter was being written, no case on net contribution clauses as such had apparently arisen, and one wonders how construction litigators have restrained themselves for so long. Just as with the underlying appointments, there will likely be institutional resistance, and insurer resistance may only serve to provoke substitution of team members. In July 2003 the Construction Industry Council published new forms, which will likely supersede the British Property Federation (BPF) forms, these now being ten years old. While no commentary is here provided, there is a continued persistence in use of provisions which will or may provoke institutional resistance, such as those mentioned above, albeit modified, and exclusion of the effect of the Contracts (Rights of Third Parties) Act 1999 also persists. Indeed, assignment under these forms is also limited to one occasion only. The conflict is thus not yet resolved, but enough has been said in these pages to suggest institutional and tenants' responses.

# 8 When it all goes wrong
## Or: Fallback in action

For many in property the 1990s' recession was a disaster in their professional and personal lives, and indeed many have not reclaimed the successes they previously enjoyed. Whilst history is said to repeat itself, it rarely does so in precisely the same way. Those who have not yet retired will still recall the boom and bust, first, of the residential property market in the 1970s, followed by the secondary banking crisis which in turn bore down on commercial property. There was a minor recession in 1981 and so on. We can leave the economic historians to chart the courses, but it is a sobering thought that as the 1980s drew to a close an overstretched banking sector relied in the main on people with very limited experience of the property lending market, and who had come to rely on a market which seemed to some capable of delivering an endless and insatiable demand for new buildings. By the beginning of 1989, with nearly two decades of qualified legal practice under his belt, it appeared even to this author that something had to give. He even dared to suggest to professional colleagues at a partnership conference that within a year they could be seeing redundancies, so receiving the inevitable response, and meanwhile the lemmings of the property industry careered on. The wider economic outlook upon which, as with everything else in this life, property depended saw gathering clouds and by early 1990 the scene was set for a fundamental change. Recession!

One deal in particular sticks in the mind, and essentially sums up all that was wrong in people's approach to development and, in particular, its funding. A certain major firm of London surveyors, well known in the institutional market, was advising a large financial institution. The sort of deal they negotiated was one which had been replicated over and over again since the early 1980s. Here was a spec-build office in the Oxford area, for which there was a perceived market. Nothing odd about Oxford, of course. The scheme was to be forward funded using conventional techniques, with the advantage to the developer on this particular occasion of a profit erosion formula: that is to say, that the developer would not be required to pay an account charge, or notional interest, on advances during the development period but that it would be allowed to roll up against advances without outlay for the developer unless the budget was exceeded. It was not.

Thereafter, until let, and this is of the essence of profit erosion, notional income would continue to roll up and be set against the developer's formula-based remuneration in due course.

Already the developer could see that, as the negotiation reached its fruition, the funding terms were beginning to look generous. The fund was beginning to look imprudent, too, but gave itself the benefit of the doubt, or rather its surveyors did. The deal was signed in early April of that year.

Of course, as with most forward funded agreements, the developer's remuneration depended upon minimum rents being achieved. Within a year, the rental levels necessary to sustain the development could no longer be achieved, and such market as there was evaporated. As the UK recession worsened, thousands of jobs were lost from the economy at large, buildings were vacated and projects foundered. Developments reliant on bank finance could no longer be sustained as interest rates rose inexorably. Insolvency practitioners prospered as they learned to make use of the instruments made available to them by the Insolvency Act 1986, and where developments had been carried out on the back of development agreements with landowners, the landowners too had to make hard choices. Often, it would be a case of agreeing a fall-back of some kind, to rescue the project or as much of it as possible, for simply to stand on ceremony would be bound to see the project collapse, and with it the landowner's aspirations for his particular interest in the built outcome.

By this time in the history of property, despite the economic folly, the legal instruments in use for development were sophisticated enough. As between employers of construction services and the professionals and contractors there was also growing sophistication, albeit perhaps not as much sophistication in the scale of things as today, given so many changes in the law in recent years.

What characterised the 1990s' recession, in terms of property development, was the sheer helplessness of those engaged in construction once employers began to see their sources of finance and remuneration drying up, and their supporters melt away. Although a new dawn was to emerge by the mid-1990s, it was to be preceded by wholesale insolvencies and collapses both within the construction industry and among related professional firms.

If no-one took risks in business, there would be no business. However, it is one thing to take risks, but quite another to fail to identify them and assess the prospective impact. The closer one is to the demand for the product and the true cost of delivering it, the closer one is to the nature of the risk. The rate for the job may also be fine, but one has earned no success if there is no prospect of it being paid, and there precisely lies the fault line. If you are a contractor who must also employ sub-contractors the point is painfully underlined. In this chapter the need for financial prudence is emphasised over and again, in the hope that sometimes needless losses may be avoided. Even the most experienced professionals cannot avoid mistakes in client relationships, and assumptions made as to the worth and substance of their clients have turned out to be false and sometimes entirely misconceived.

Larger contractors and professional practices have finance departments led by suitably qualified professionals: there is a useful further role for them beyond the accounts of the business in assessing the financial strength of third parties with whom one is dealing. Those less fortunate, smaller organisations and professional firms, are perhaps the more vulnerable, perhaps simply because they are led and peopled by those having particular professional and service skills, but who are not immediately touched by such considerations. Neither is it sufficient to assume that by delivering the detail of a proposal into the hands of one's solicitors, one has taken all the necessary safeguards. The practical and harsh reality is that sometimes those involved with the immediate issues surrounding the legal instructions may be little better equipped: the less so again as some law firms so compartmentalise their respective specialisms as to isolate them from underlying commercial considerations and from those more attuned to their impact. Their job, for example, is the detail of construction contracts, and when that is done, there it lies. Not that it is never the client's fault, whoever he may be; instructions given on a need-to-know basis are in turn the most difficult to implement. As a lawyer perhaps one should not be too ungrateful, after all, as the negligence specialists must earn their crust too. There is money in folly and a new culture of awareness is required.

### Pre-lets

As one example, one of the lessons of the recession, and thus one of the consequences, was that if property finance depended on successful letting, the safest way to achieve that was to meet actual demand as opposed to perceived demand. The sobriety now enveloping the bankers compelled them to realise that speculative development meant speculative reward. The notion of high risk, high reward, is well understood by experienced developers. Pity the bankers who saw experienced developers taking risks as somehow isolating them from exposure. If the object of bank lending is to be repaid with interest, or at all, then the conditions must exist for that to be achieved. The same sentiments apply to recovery of professional fees, and payment for construction. Elsewhere in this book we have talked about covenant-led development as opposed to deal-led development, particularly in the field of forward funding.

Somewhere in all of this has to be the financial strength to support the project or, in the case of institutional investment, its life blood, income. The creation of value by securing income has at last been grasped by the bankers. But lawyers may also take comfortable satisfaction from the inevitable, that human frailty will soon lead the bankers to forget: the only question is when. Meanwhile, the world must be viewed in a certain way.

Consequently, having entered the twenty-first century we now find that unless a building is prescribed for owner occupation, tenant covenant is largely the key to it all. If the employer is not employing from own resources, then his source of finance is of paramount concern. Construction professionals, and contractors in particular, must now be circumspect – they after all are shouldering the ultimate burden of the massive capital cost before determining their profit.

Not that they were never circumspect before, but the money trail to the source of finance, and the accessibility of cash, must be examined with great care. Think of what happens if the tenant is afforded the opportunity to walk away (see below). As an employer the SPV, single-purpose vehicle, may be an ideal medium for banking purposes, affording the kinds of safeguard we have discussed to the banker and to all the other players behind the developer but, in its own right, it is an unworthy employer. It sets out to have limited resources which are primarily available only to others in an insolvency. It also sets out to keep its begetters invulnerable so far as may be from attack by others. However, just as the SPV will be guaranteed in all its dealings to suit the banker, the underlying employer who refuses to commit to his construction and professionals should remember that, if the SPV should become insolvent during the course of the development, the directors themselves may become vulnerable at the instance of a liquidator for trading while insolvent, in particular through engaging in professional appointments and construction contracts which cannot be supported. Granted that through our insolvency laws the full range of remedies is not immediately available to all, invocation of powers against directors by liquidators tending to be used sparingly, there is much to be said for endeavouring to instil good business practice to compensate. For some construction professionals this will be a novelty.

Remember also that the terms of a pre-let agreement go to the heart of the viability of a project. The building contract may indeed include in its employer's requirements, the development requirements of the tenant which are vital to acceptance of the building, the delivery of the income stream and the satisfaction of banking requirements. Look a little more closely, therefore. If the agreement for lease prescribes an event of default in the event of insolvency of the developer (SPV), such an event is not only inherently unbankable because the tenant will thus be afforded the right to walk away, but that right to walk away means, equally, the dessication of finance and exposure of the contractor to loss of remuneration. Developer breach, such as failure to deliver on time is another matter, and may still have the same consequences although funders will usually be quick, in their own interests, to ring-fence their exposure.

Development agreements, particularly developer/tenant agreements, are important documents and cultural changes would go far to elevation of contractors and professionals to the status of principal players, in accessing underlying contractual arrangements as part and parcel of the fabric of negotiation. At the present, and with no apologies for shattering the illusion, they are often accessories and no more, however closely they think they are working with the 'client'. If they are prudent business people, they should be on enquiry (and to be fair most are), and while they may fear the consequences of enquiring too far, that must be their judgement. Of course, in numerous cases, developers and contractors are working as joint-venture partners anyway, but it is where the construction side of the equation is wholly at arm's length that a culture of co-operation in a single aim needs to be engendered. The insurance industry, in its own way, is moving in the same direction – see Chapter 9. However great the lip service, the 1990s' recession bears witness to its consequences. At least as a starting point, a well-constructed pre-let can also provide a comfort zone for all concerned, including those involved in construction.

## If it does go wrong, what then?

Whilst economic down-turn and market conditions will exact their own revenge, the possibilities for failure are endless. Failure brings into focus the safeguards which the players put into place, or thought they had put in place, or did not think to do so. The underlying purpose of this chapter is, therefore, to convert hindsight into foresight. In so doing, it adds a dimension to the principles behind the legal aspects of procurement.

Who is the employer? We have already identified the prospective nature of the employer as being anything but the substantial entity we would hope it to be. Chances are that when one is being invited to tender, or one is seeking the opportunity to tender, one may indeed be confronted, initially, with a substantial name. Even though much work may be done on contract preliminaries, in the case of a design-and-build contract followed by a full raft of contractor's proposals in response to the employer's requirements, the true identity of the 'employer' may emerge only at the last moment, when draft documents are offered, in the form of an SPV development company. Now is the time to put in place safeguards which will lead to available fallbacks.

Accordingly, part of the cultural change should be to place oneself on enquiry on a number of fronts. The existence of a pre-let, in the absence of construction for owner occupation, is the first layer of comfort. If the enabling agreement for lease with developer's obligations finds its way into the contract spec, it is helpful to look further than the developer's obligations which the contractor and professionals must heed in the performance of their respective duties. What about the tenant itself? Does it provide a good institutional covenant? If not, does it have a guarantor? Not that the most positive of answers to the above questions will add to financial safeguards in the immediate sense, but the knowledge that the end-product will constitute a valuable investment puts the other parties and their respective roles into perspective of a certain kind. It is when the financial fortunes of the tenant founder that the principal players, fronted by the immediate employer, begin to safeguard their own positions and avoid as much exposure as possible.

If an SPV employer company is not resourced, crying foul will not make it better. If the SPV is the creature of a substantial entity, and even if that entity (or entities) has not provided a guarantee, the pursuit of legal action for recovery of unpaid sums is inherently difficult albeit not necessarily in vain. First, the underlying principals' ability to do business, and their reputation in the market place, may be impaired if they are seen to use what is otherwise a convenient banking mechanism to avoid liability. Where insolvency looms, and if there is the slightest hint of trading while insolvent (that is to say, of incurring debt without the resources or the material prospect of acquiring resources to support it), then is the time to cry foul to the bankers and financiers and seek their help in expanding the recourse as remotely as the law allows whether by tracing, as it is called, or by calling directors personally to account, so far as available to a liquidator according to statute.

Where lenders and investors are concerned, an insolvency event triggers inevitable consequences under related development and loan/finance agreements. Moreover, where the project is founded on a development agreement with a landowner, an insolvency event also triggers an event of default under that agreement and so sets in train a cascade of events of default. Those concerned with financing the development must take defensive action on their own account, but in so doing, by ring-fencing their exposure, they limit the prospective recourse of others, but that is not to say that they will not be interested in the cause of default or in maximising recovery.

Much of the procurement process, as prescribed by development agreements, and apart from the construction and professional elements themselves, paves the way for the consequences of failure. The more circumspect one is, the closer one should be to the principal players, and their individual constitutions, and to the way they relate contractually to each other. Starting with the landowner, if he is not also the employer, his landowner/development agreement essentially affords the developer licence to be on-site, so even if the developer has the benefit of a parallel lease, the prospect is that it will also be expressly determinable in the event of default under the development agreement. If he cannot pay the bills and work stops his entitlement to occupy the site will cease with it, and so also the ability to proceed with the construction. That much, at least, is an irresistible process and few crumbs of comfort can be drawn there. Moreover, do not expect to negotiate a guarantee from the landowner for the performance of his developer (see below as to guarantees), although there will be times when his interests are so critical it may actually be in his interests to provide one.

It is of the essence of landowner/developer agreements that, in the main, the developer is to carry out the development at his own expense and in some way to provide benefit to the landowner whether through payment for the site in a variety of ways, including overage, or by affording the landowner some benefit or interest in the built outcome. Again, do not expect lenders or indeed investors, to guarantee the developer's obligations. Their interests lie in protecting their savers' deposits. They will have been busy enough in requiring security of various kinds including, perhaps, from the contractor itself by way of performance bond or parent company guarantee. Remember that a lender is in the business of lending for profit and having his loan repaid. The most effective way of doing this is to pre-package the development so that, if required, he can pass it on to others who will carry on and complete the development. The developer's lender (or his onward buyer) may then become your best friend, stepping into the shoes of the developer and operating the fall-back provisions of the development agreement, and by so preserving his own security allowing the project to continue through its insolvency representatives or by a new developer.

When things are going wrong, however, an investment purchaser from the developer should be regarded rather differently. He is the end-buyer and if the investor is also making advances under a forward funding agreement, it is not usually a matter of disposal but of eliminating the developer and choosing whether to carry out the development itself, which is not usually a difficult decision if the development is substantially advanced. Where a development is forward funded, and unless the construction team are themselves at fault, their best bet is to work closely with the fund who will, in any event, have been represented throughout by the surveyor charged not only with supervising the work but certifying draw-downs. But, again, do not expect a guarantee. Be thankful to be novated under the terms of a warranty deed. Substitution provisions are potentially another best friend.

If the employer is an SPV, it makes sense to seek an employer's guarantor. Consistently on experience, the subject is rarely broached, and one wonders why. It perhaps comes back to delegation of detail to people who are not entirely at the heart of commercial decision-making. Further, it will likely not be in the imagination of trade associations and insurance representatives who will struggle enough with contractual provisions themselves to avoid claims against their members' policies. Neither will it necessarily be in the imagination of some less-experienced members of the over specialised departments of some law firms. Someone has to see the bigger picture. One hopes that hereafter, prompted by this book, professionals and contractors will, as a matter of course, consider the commercial risks and where appropriate seek guarantors of SPV employers across the board. If the offer to tender came from the top company, it should put its money on the table. It is not a complete answer, of course, as large companies are no less vulnerable to insolvency in a downturn. What is inexcusable, however, is taking a contract from an SPV company, or from any person or body with limited assets, knowing that it has no assets of its own, and without even considering the downsides or attempting to improve on that position.

## Third-party commitments and warranties

Until it became clear as a matter of law that a variety of devices to link third parties, such as funders and tenants, to members of a construction team were bound to fail, the principal contractual relationship would be with the employer alone. Pre-1986, of course, an SPV developer was fairly rare. More likely, one would find property companies with development and investment subsidiaries, each with fairly powerful covenants in their own right. Where that covenant was lacking, perhaps when banking or investment depended on it, a parent or associated company guarantee might be offered. That apart, the employer of construction services stood a chance of being rather more substantial then than the necessarily nominal nature of an SPV employer.

When an employer's ability to pay dries up, steps must be taken immediately. Even though it appears a hopeless cause, the default provisions of the contract (for the purpose of this discussion including any appointment) must be reviewed. Institutional influence will likely already have eradicated the imposition of conditions on the employer, if the employer has not anticipated therein his negotiations with the contractors and professionals, such as appear in the professional bodies' standard forms. Even the most poorly drawn contract is unlikely, on analysis, to require the appointee to continue to perform in the face of prospective non-payment and existing arrears. If the figures are, or are prospectively, substantial, then legal advice should be sought straightaway. Even if the personal relations have been good and apparently remain so, it is all too easy to hope that something will turn up. It is time to look to the letter of the contract. Never rely on something simply turning up.

On the development side of the fence, similar considerations apply particularly in relation to joint ventures. Failure to instruct lawyers tends to arise from not realising the nature of a business relationship and the need to express it. When it founders, only then do the players begin to realise that whatever they believed their relationship to be, its resultant legal definition is far removed from their perception of what it might have been. One still comes across joint-venture situations which were not only not in consequence fully documented, or sometimes not documented at all, and the commercial terms were by no means clear, but when the hoped-for benefits did not materialise the weakest member complained that somehow he was robbed or duped. Construction professionals who embark on development (as opposed to construction) in concert with others with whom perhaps they have worked for years and have come to trust, can readily lose all by failing to define their relationship and by failing to obtain the appropriate legal advice. A book on joint ventures should be regarded as a good companion.

When difficulty knocks on the door, by then it may be too late. However, on the building side and with construction contracts of their various kinds, there is greater precision these days, and corresponding legal input, and the strengths and weaknesses will soon be found in case of default or dispute. Failure to pay, and an absence of assets, will not be cured by the best legal drafting, but insofar as there would be a remedy it will not be helped by inadequate documentation. Seek advice and ensure that all appropriate demands are made. Advice is also necessary upon entering into an appointment to ward against steps being taken by the appointee that actually worsens his position. It is extremely awkward to find someone trying to amend a draft document to his actual detriment.

Neither is drastic action, taking the law into one's own hands, to be condoned in any way, shape or form. It will earn judicial displeasure and almost certainly make matters worse. It is time for further anecdote: many years ago, a developer client had negotiated over a very short period for a contractor, which was a major name in house building, to erect an office building in part of what were then the rural outskirts of London. Today it is a village of office buildings! A meeting was held on the Monday morning with lawyers present (who had only learned of the matter and of its apparent urgency the previous Friday afternoon) to finalise the heads of terms: the decision was made to proceed under something perhaps less detailed than a JCT design and build, and to work on the basis of a development agreement between developer and contractor, which in its negotiation and indeed final form was at least adequate for the particular task, as indeed it proved. In the afternoon everyone sat down around the table, this author with his dictating machine at the ready, clauses to be dictated and sent to the secretaries as each section was completed. By 7 pm that night, the deal was done and the document signed. The development proceeded well, the detail of the documentation was entirely adequate, albeit slightly unconventional, to reflect the heads of terms. A series of stage payments had been devised in conventional form, and all went well until the last principal payment. There was a dispute but the developer held firm. The building was practically complete, and a security guard was installed. The first one knew of a problem came in a phone call from the client at lunchtime on a Saturday. The managing director of the contractor had taken some thirty heavies armed with pickaxe handles down to the site. They gained entry to the building, threw out the security guard (he in turn calling the client), and proceeded to brick up the ground floor windows and exterior doors. We were in court on Monday for an injunction. As the client conceded, stopping the payment, whilst hardly for no reason at all, was yet on shaky ground, but the conduct of the contractor so angered the trial judge later as to more than tip the balance. Don't do it!

## Other kinds of principal players

For convenience, most of this book has been written around the presence of an SPV entity as the developer, on the basis that where other players are to be involved, whether as joint-venture partners, or as principals in their own right, this will be the likely form the developer will take.

Where development is carried out for own use, you take the client as you find it, and one works from there. It might be, say, a housing association which, albeit having 'Limited' after its name is, actually, an industrial and provident society. Or it might be a limited partnership, already discussed, hardly a company at all but still having 'Limited' after its name. It might be a partnership, or it might be a limited liability partnership, an LLP. Or it might be a trust.

The investor who employed the developer may be acting as a trustee, or trustees, and if the development is being carried out directly for a trust, caution is required by both sides. Trustees are custodians of trust property, and should properly be subject to the terms of a formal trust deed. Under a looser arrangement, an unincorporated association, albeit unlikely to be found in the case of substantial development, the named owners will be trustees of land under property law rules.

Trustees are vulnerable people however their status has derived. In the absence of contractual safeguards they have personal and unlimited liability. The first expedient is for their appointee, contractor or professional, to ensure that when there are joint employers, they are stated to be jointly and severally liable. (Then, to put the chain of command beyond doubt, the validity of instructions needs to be made clear so that, say, an instruction is only valid if given by both together – if there are only two of them, say – or by one of them or someone nominated on their behalf.) The simple expedient of making them jointly and severally liable is that individually they can be pursued for the whole of the indebtedness. That is the first safeguard.

However, trustees are vulnerable in other ways in as much as, quite wrongly, it sometimes happens that they believe that their liability is confined by their function. This is patently not so in law, and trustees properly advised will, in any contractual situation, include a declaration that their liability shall be limited to the trust assets. If you look at a competently drawn lease or an agreement for lease, and even if the trustees are the landlord – let alone the tenant – that is what you should expect to find, and if not then the trustees were not properly advised. The position is no less so if the landlord is a financial institution in the guise of a custodian trustee. So also should it be when they enter into contractual commitments with those providing professional and construction services. If the declaration is not present, and assuming that they as employers are expressly liable jointly and severally, so easily done at the stroke of a pen, then they can be pursued individually to the full extent of the employer's indebtedness.

Now look at it the other way round. Expect a professional partnership to be appointed on the basis of joint and several liability, and while one should try to ring-fence that with limitations of insurance and net contribution clauses, as has been seen, the prospects of ring-fencing are limited when commercial banks, institutions and substantial tenants are involved.

## Recourse and preference

Enough has therefore been said about the coincidental benefits to a developer of using an SPV company, thereby precluding or limiting access to the underlying developer and other principal players. However, where there are funding partners, although one may not be able to sue them directly, one should try to remain as close as possible to them and their representatives throughout: one day, simply knowing who to talk to may be useful. If the provisions of forward funding agreements and of loan agreements facilitate, as they often do, direct payment by the funding source to the contractor and professionals, the sooner that funding source knows of a difficulty with the developer the better (see also above as to trading while insolvent). Whilst developers will try to negotiate their funding documents so that draw-downs are routed through them, many funds and lenders insist upon retaining the right to make payment direct. In other words, at the first sight of difficulty, that difficulty may lie with the developer and it may be that the funders will be keen to keep the funds channelled to the appropriate recipients, so talk to them.

At that point, one may still be well short of developer insolvency. A spat between a developer and the fund's representative is by no means uncommon. Find out who your friends are, and make new friends if you can. So long as the funding source has cause to see the project continue, that is one's best protection. If the line appears to be that the developer's default may be sourced to you, take legal advice immediately. However, one does not necessarily have the comfort of banks and institutions in attendance, and other safeguards must be taken. Much of the agony of financial failure can be avoided by simply understanding the nature and financial strength of the employer. SPV companies apart, and the relationship with holding companies, other assumptions must be laid to rest as well. The recent case of one of England's cathedrals running out of money through a failed project is a case in point: at the point of writing, the Church of England itself has not apparently flown to the rescue so, once again, we have a case of appointees perhaps making assumptions about the source of payment which, had they thought about it, could never have been reached. However, to the extent only of media-filtered anecdote, the prognosis has a familiar ring. The same question always arises, what is the financial strength of the employer and what is the route to recourse?

An insolvency practitioner, marshalling the assets of a person or body who has failed to pay debts and in so doing has triggered an insolvency event, has a prospectively difficult task. Hopefully, the fruits of his labours will be a statement of affairs identifying assets and also showing that they exceed liabilities. Even if liabilities exceed the assets, then under the rules of priority there may still be sufficient assets for the unsecured creditors, who unhappily are always at the bottom of the pile, to be paid a dividend, however much less than the whole of the indebtedness.

Insolvency brings with it a number of unhappy side affects, one of which is 'preference'. Once upon a time it was more realistically termed 'fraudulent preference'. It is beyond the scope of this book to pursue the detail and, again, specific legal advice should be sought. However, it should not be thought that because one has had the good fortune to be paid, but the next man has not, one has somehow got away with it. So, say that at a particular moment a company has £2 million worth of debt and £1 million of liquid assets at a time when it was clearly trading while insolvent. If it pays one set of creditors in full, or in proportion more than the others, the scene is set for the prospect of recovery and redistribution. It is a bit like the trick of putting one's house in the wife's name to escape creditors, effective if it was done long enough ago in law to put the assets beyond reach, but otherwise if done more recently there may be recovery through the process known as tracing. It comes down to timing.

## Guarantees and indemnities

As part of the fabric of building in protection at an early stage, and assuming the employer has conceded the nominal status of its SPV and is willing to lend some covenant strength, how is this actually done? The first thing is not to attempt it yourself and to get legal advice in drafting the documentation.

The terms 'guarantee', 'indemnity' and 'warranty' are so loosely used in common language as to imply some commonality of effect. In law, this is simply not so. If I go into a shop and buy some electrical goods, I will doubtless come out not only with the goods but with a piece of paper calling itself a guarantee. It may be issued by the retailer or it may be available from the manufacturer (or I may have both). Of course, the law relating to sale of goods has rules of its own, but a retailer's guarantee is no 'guarantee' at all. It is, actually, all part of the same contract under which, whatever the terms, the retailer actually warrants the quality of the goods, undertaking also, say, to replace or repair if found to be defective through manufacture in the first year, or whatever. A manufacturer's guarantee is essentially no 'guarantee' either, which does not necessarily mean that it is unenforceable, but simply that it has been mis-described.

A true guarantee is a contract in its own right, whether given for consideration or executed, under which the person providing the guarantee, the guarantor, guarantees the performance of a duty owed by a third party. A guarantee may say that the third party will perform the duty in question and provide that the guarantor will, in case of failure, perform that duty itself. So, the essence of a guarantee is to underpin performance by a third party, and if the guarantee is called and the guarantor performs, he in turn has an implied right in law to be indemnified by the original defaulting party. An indemnity thus constitutes the making of recompense.

As between the guarantor and the person to whom he has given a guarantee, there will usually be allied with it in the same document, indeed in the same clause, an express indemnity in favour of the beneficiary of the guarantee. Often the guarantee will say that the guarantor will perform, or otherwise indemnify, etc. An indemnity being a contract to provide compensation, in other words damages, for the non-performance of the particular contract which the indemnity underpins, this is also the alternate route if performance of a guarantee as such is the less practical.

Guarantees and indemnities are essentially third-party contracts. So, a purported guarantee or warranty given, say by the retailer to his customer, is simply a term of that contract and no more. Contrast a construction 'warranty' which is given by a professional or contractor, not to his employer, but to a third party. That third party, unless it has express legal step-in rights, or is separately novated, or has taken (with consent where required) an assignment of the benefit of the contract in question, has an interest in the built outcome but is separately identified from the employer. The warranty may not govern just skill and care, but may also provide for actual performance in accordance with its terms. The benefit that the third party may derive is limited in law by the extent of his interest, and it is of course well established that if or insofar as he in turn disposes of that interest, then the benefit he has will diminish accordingly.

## The value of assignments and warranties

However harsh the terms of a construction warranty, in terms of the exposure for which it prescribes, and where provision is included for instructions to be given by the third party in question in place of the original party to the contract, say a bank or fund, or in case of assignment of that warranty (or of the original contract), it has the potential of great value not only to its beneficiary but also to the warrantor. When such provisions appear, they should be understood and distinguished.

A bank lender, realising its security in repayment of a loan, will likely seek to transfer on the whole package, including the land, the project and the construction team. The contracts themselves will perhaps have been the subject of security assignments, or warranties may have been given, empowering substitution or step-in (novation). If not, one is staring into the abyss of employer default and all the financial consequences that go with it, and so a development loan package will include among the security instruments a raft of security including over the construction package. Alternatively, for any number of reasons, the employer may wish to assign the appointment itself during the course of the contract. Let us deal briefly with appointments and contracts themselves in the context of this chapter. As mentioned earlier, there are likely to be provisions designed to give some comfort that the assignee will be someone with whom one can work and generally live. By contrast, the ability to assign a contract to a bank lender as part of a total security package, may be expected to be entirely without restriction.

Novation clauses apart, provisions for assignment are usually simply drafted, with no express provision for recourse against the assignee. As a general principle, when the benefit of a contract is assigned, indeed any kind of contract, is assigned, and assuming that it is assignable and is not deemed in law or otherwise expressed to be personal, it is precisely the benefit and the benefit alone that passes, but while there are burdens, they remain enforceable while the assignee has the benefit. The burden of outstanding liability therefore must be dealt with separately and provision must be made for a direct obligation between the assignee and the contractor.

So, say the contract has been assigned, the original employer is otherwise still an available recourse in addition to the current employer, unless he has been expressly released. If the employer was supported by a guarantor, say its holding company, now is the time to look at the terms of the guarantee. The nuances of the guarantee clause will now be directly in point. What if the original employer has been liquidated, and what are the insolvency consequences? The wording of the guarantee is vital because when liability dies upon eventual winding up of the SPV, so does the guarantee unless it is properly drawn to underpin a variety of consequences. It is not so much legal mumbo-jumbo, it can be a meal ticket delivering either three courses with all the trimmings, a bowl of gruel or just hard cheese.

It is also a simple matter to require contractually that notwithstanding an assignment of the benefit of the contract, the professional or contractor in question shall have no duty to the assignee save upon the assignee giving a direct covenant to observe and perform the employers obligations. Hopefully the obligation will be written to include any outstanding obligations on the part of the employer, and not just future ones.

Where warranties are given, which survive completion of construction, and particularly where the person having the benefit of the warranty has not adopted the role of employer, the scope for corresponding obligation is limited. Rather, it is more likely that the benefit instead has potential for limitation. Thus, the benefit of a warranty in the hands of a tenant, say, may be limited by the consequences of the tenant's own actions, either in course of use of the property or because the tenant carries out works of its own which compromise the beneficial effect. One may rely upon the ordinary law of contract to define that limitation, but if there are express issues known about in advance, one can avoid losing the argument because of that knowledge by making an express contractual distinction. In the course of construction, however, some warranties imply the prospect of novation: that is to say, that even if the contract is not expressly assigned, machinery will exist for direct instructions to be given and accepted, in which case it is essential that provision also be made obliging the person giving those future instructions to undertake the employer's obligations. Instead of an assignment, what is happening is the creation of a new contract (hence the need for the employer to be joined as a party to give his consent) between new parties which thus requires these obligations to be restated (and for clarity the original employer may be discharged save, perhaps, for outstanding obligations).

## Other difficulties

Default and disputes apart, a contract, for the purposes of this continuing discussion always including an appointment, may founder for other reasons. Not least of these is the intervention of events or circumstances which render either party innocently unable, in part or in whole, to perform its obligations. This can apply equally to employer or appointee. This may give rise to application of what is known as the doctrine of frustration which, if applicable in context, also means that no blame can attach, because an event has intervened over which the party in question has no control and which could not have been foreseen. So, for example, in a conveyancing case many years ago, a seller's ability to appropriate a deposit upon the buyer's failure to complete the purchase failed because the available funds to complete the purchase were unilaterally frozen in a bank in Nigeria, in circumstances where the buyer was entirely blameless.

When things go wrong in the contract, the apparently defaulting party should beware pleading 'Not my fault, guv', as his only line of defence. The doctrine of notice is a powerful antidote to legal 'frustration'. The law does not sympathise with carelessness or a cavalier attitude. It behoves all the players to be on their guard to ensure that the way is clear for them to perform their respective obligations. In the case of construction in particular, just as a buyer of land is on enquiry and is deemed on notice of those things which one could reasonably be expected to know, particularly if a matter of public record, so also does this reach down to the players in the construction process. There is a mass of information available much of which construction professionals will seek out for themselves, and other aspects of which they will seek from their prospective employer, he in turn having made due and proper enquiry (if he is careless in turn, and misleads his appointee, that may serve to assist or even exonerate the appointee, but do not rely on it). That duty will, in practical terms, often be performed by the employer engaging other professionals initially, surveyors and solicitors in particular, on that fact-finding exercise. It always seems to be the solicitor's fault, doesn't it, and so often unhappily that is so.

But let us assume that all of that is done and that the architect, say, and the engineer also, have called for all of the available information to be fed through. Is there something missing, perhaps, some vital element which his professional training and knowledge should tell him requires attention? It is too easy to generalise, and thus be unhelpful, but the central point for every person engaged under a contract is whether or not any failure to perform on his part is due to frustration, which is blameless, or is in spite of actual or deemed notice, which is culpable. If there are specific issues of concern, one should never trust to luck, they should be aired and resolved, and if necessary special provision should be made for absolution from responsibility (but contrast general exclusion from liability which may not escape a finding of negligence). Of course, professional judgement is important too, and this or that view may have to be taken. Take it carefully, or negligence looms. If only solicitors could enjoy limitations so easily: they, by contrast, have responsibility for anything and everything which comes within the ambit of their instructions, and absolution by the client may still leave them open to a charge of negligence if the judgement was one that could only be made with the benefit of the solicitor's advice. In a recent incident, this author's practice was instructed not to advise on the construction aspects of a recently completed investment: in a nutshell, the nearest one could get to absolution was to advise that if his practice was not to be so instructed, then others should. On the strength of that, several boxes of construction documents had already arrived from the seller's solicitors, and were passed to the investors' building surveyors, but travelled no further. Enough said.

The extremities of responsibilities suffered by lawyers apart, it is clear that so much of what goes wrong in construction is common to all commercial contracts. It is easy for lawyers to pontificate about the rights and wrongs, because they are the ones who pick up the pieces. They make a living out of constructing contracts so as to limit so far as possible the exposure of their clients, and from picking up the pieces where that process fails. Solicitors will themselves be negligent if they fail to use their imagination, but the dangers for any professional person taking instructions on a need-to-know basis are only too clear, if not to anyone else, then to solicitors. There is a clear need for everyone involved in the process to look far beyond the task which the proposed contract prescribes, and to assess honestly and carefully its achievability and the prospects of being paid for it. The 1990s' recession shows time and again that few of these considerations ever entered the minds of many of those engaged in the construction process. It is trite to say they were not commercial, but in some cases it was no more than that, and financial naivety was rife. In a buyer's market, 'Do you want the job or don't you?' may seem irresistible. To those who suggest 'You have to be realistic', one can only say 'Exactly'. We are therefore back to risk: risk is good if you are prepared to accept its consequences, failure to assess it is foolish.

Of course, much risk can be insured, and aspects of risk management are addressed in the next chapter. But whatever the subject matter, insurers are not in the business of wagers or bets. There is also a basic distinction in law. An insured risk is itself usually only accepted on the basis of due enquiry. An insurer will not insure a defective title to land, for example, unless an appropriate degree of due diligence is taken. If the consequences can be foreseen, whatever the event in mind, insurability is undermined accordingly. An insurer who fails to assess the risk adequately has only himself to blame.

## Default and disputes

The wearisome process of construction disputes hardly has a place in this chapter. A dispute is a dispute and must be resolved, and there are plenty of mechanisms for that including as introduced by recent statute. There is no need to repeat what is covered in standard works. For the purposes of this chapter, however, particularly where there is complex development with a variety of principal players, including funders and tenants, resolution of disputes is fundamental. It is fundamental because of timing, and timing goes to the root of the range of development agreements which gives rise to the developer's ability to pay for construction services. Play unfairly, and you bite the hand that feeds you. If you think that the developer is unfair, engage the attention of other principals, perhaps not the tenant initially if that might provoke him to walk away, but certainly the funder and anyone else with a financial interest in the built outcome. Be discreet, take legal advice, do not tell tales for fear of defamation, but when difficulties arise engage the attention of others concerned in a positive manner to assist in resolving the problem. If the developer is playing unfairly, he may be in default of his contract but, more particularly, he may place himself in default of his funding terms. It is better to have other players on side if you can, better still all of them, and above all take good legal advice. Good luck!

# 9 The management of risk

Or: 'Tis better to be safe than sorry

A book about development can scarcely ignore the subject of risk, risk of liability in some form, risk to the success of the construction process, risk arising from remediation, and risk arising during the life of the resultant investment.

It is of the essence of the development process that a holistic approach is desirable. As viewed by a lawyer, the containment of risk is to be approached both in a proactive and in a reactive way. The reactive way, of course, is the attribution of blame and loss and, if necessary, taking steps to mitigate it and the proactive is avoidance in the first place. It is in the nature of the proactive way that the rationale is to avoid bringing in the lawyers later. A lawyer's role, so far as it is proactive, is to ensure that as part of this process, risk is contained by judicious contractual and security measures. When these fail, either he has failed, or one or more of the players has fallen victim to those measures, or someone is breaking the rules of the game.

It is hardly surprising, therefore, that professional risk management has emerged over time. Insurance in particular has an important role to play. The prudence of insurers is critical to the prospects of a claim on a policy, and insurers, in particular, are keen to see the prospects of risk mitigated as a basis for the acceptance of risk.

As indicated in the preface to this book, help has been at hand, and so the provision of material for the preparation of this final chapter has been particularly instructive for this author in viewing the effect on development and construction contracts and the consequent impact on drafting. As every lawyer will also say of his client, it is helpful to listen to someone else for a change, and this chapter is the result.

Incidentally, with effect from 14 January 2005 the Financial Services Authority (FSA) assumes responsibility for the regulation of general insurance, following the Insurance Mediation Directive published in January 2003. 'It is important for anyone involved in property management and development to understand the basics of this legislation as it could well affect their relationship with tenants and other interested parties' (Aon Property Eye, Summer 2004). The dead hand of regulation has fallen again on the insurance industry and those involved with contracts of insurance including in relation to development. As always, there are qualifications but the purpose here is simply to underline that a further tier of regulation has arrived.

## Anecdote and attitude

The containment of risk, and the relief afforded by insurance, is an evolutionary process. A constant factor is the interruption of business. Of course insurance is no substitute for risk management. Indeed, as risks become more expensive to insure, for a number of reasons, containment is the more important. Moreover, however well defined a risk, if the insurer itself is weak, relief may yet not be at hand.

Every now and then a major disaster somewhere in the world provokes thought and change. Piper Alpha and the World Trade Center come to mind. '9/11' has thrown up some interesting statistics, against its particular background and of course the damage was not accidental, but deliberate. Whilst aviation losses amounted to only 10 per cent the towers themselves 9 per cent and other property 15 per cent, the biggest single loss was business interruption, 27 per cent. Insurers have thought hard, going 'back to basics', narrowing the extent of cover, and with higher premiums driven in turn by the re-insurers, thus reducing profits. In the interests of managing the risk, the focus is ever more on design and protection.

The UK Loss Prevention Council design guide is thus aimed at property/business interruption, rather than personal protection, and had indeed adopted this focus prior to 9/11. Perhaps with existing ideas of defects liability insurance in mind, insurers' risk engineers require to be involved during concept and design stages. Retrospective compliance is expensive (see below as to defects liability insurance). Awareness has not always been enlightened, however. For example, the dangers of combustible composite panels have been known for a long time, but the insurance market has been slow to react to it and recommendations were not enforced. These are now perceived as scarcely insurable. The discouragement of the use of deleterious materials, in construction contracts and development agreements, is thus more than appropriate practice. The consequences of inadequate risk management are significant: the cost of remedial work may be prohibitive, premises may remain unlettable and vacant, the income stream may therefore be reduced or dry up, and institutional quality will be impaired. This being so, someone's pension may also be affected, and so on.

Since 9/11, the insurance market has been further squeezed by liquidity issues. Insurers have reduced their exposure to volatile and complex risks and, necessarily at a price, there has been a 'flight to quality' both on the part of insurers and insured. In turn, governments are under pressure as insurers of last resort (see below as to 'Pool Re').

As another example, whether engaged on the acquisition of land for occupation or development, the lawyer is increasing hedged around by the prospects of responsibility for failure to make appropriate environmental enquiries. The planning process itself is, of course, subject to input by the Environment Agency. Climatic changes have resulted in serious flood damage now being considered as 'normal', and following the floods in this country of October 2000, ABI (Association of British Insurers) insurers have reacted in turn, bearing down on high-risk areas where cover is now increasingly difficult or expensive to provide. ABI insurers will work with policyholders to see if the risk can be improved, particularly where improvements are made to the property to guard against risk of flooding. There is also a political dimension, as the risk can often be improved by increased government expenditure on related infrastructure.

Construction insurance has become more expensive, with consequential pressure on small contractors who cannot work without public liability or employers' liability cover. The cost of professional indemnity insurance across the professions (not least solicitors) has increased massively.

The scale and scope of insurance risk is best left for the experts. By way of more serious anecdote, for example, toxic mould was identified in the early nineteenth century and its dangers were known by the 1920s. It presents health hazards to inhabitants and others. It is understood that much can be attributed to the improper handling of water damage, and to the omission of prompt dry-out. It is apparently a major source of claims in the United States, which are continuing to rise. It is noted that the ACE professional indemnity insurers are now excluding toxic mould from the renewal cover, and this and related exclusions, regardless of the source of the mould, have the prospect of appearing more generally: an unattributed source has commented 'mold is now a transaction disclosure issue in the US – big time!'

There will be more on terrorism below, but suffice it to say in this section that post-9/11, most primary policies exclude terrorism entirely for property damage and business interruption. Other kinds of insurance will have limited protection. Policy exclusions have developed in line, reflecting changes in the likely sources of attack, an excluded act now being 'the act of any person or groups of persons whether acting alone or on behalf of or in connection with any organisation or government committed for political, religious, ideological or similar purposes including the intention to influence any govern-ment or put the public or any sections of the public in fear'.

While exclusions proliferate, so also are new risks identified, and with them yet more exclusions. These may cover, for example, digital/cyber risks, virus or similar mechanisms, hacking and denial of service (network/systems). Older forms of exclusion remain, e.g., those associated with war, riot, civil commotion, nuclear risk etc. A whole raft of changes has thus been evolving and, in deference to the experts, one is advised that any detailed comment is unlikely to stand the test of time.

The ACE scheme for PI cover excludes, in relation to terrorism 'any consequence whatsoever resulting directly or indirectly from or in con-nection with terrorism regardless of any other contributory cause or event' including 'any action taken in controlling, preventing, suppressing or in any way relating to' terrorism. Whatever the prospects, watch out for change.

Professional indemnity cover is becoming more difficult for contractors having design responsibilities, with a perceived move to an aggregate basis rather than each and every loss. Further pressures include an indemnity limit excluding legal costs, and higher excesses. From the lawyer's perspective, this is drastic indeed: if incurring legal costs can contain the consequences of risk, so much the better. Clauses in professional appointments and warranties, often inserted by appointees' representatives, may seek to exclude liability of individual partners and employees, in reaction to difficulties in insuring them individually.

A changing market is particularly important in the context of the insurance provisions of collateral warranties, or indeed of the underlying appointments themselves. Assuming the appointment is executed and thus has a limitation period of twelve years, there is an obligation to maintain PI cover at a certain level during that period. Most employers will provide at the outset for some future relaxation in these levels based upon what is available in the market at commercially reasonable rates. Employers will be wise to consider how far, and even though the legal drafting may be impeccable, such provisions may continue to assist and whether change in circumstances may suggest an alternative approach. From the point of view of the insured under a PI insurance, it is not enough to think in terms of 'standard' cover and to concentrate perhaps solely on attempts to reduce premiums, in effect competing for the insurer's capital, i.e. the resources available to meet claims. Insurers in turn will take more care, discriminating heavily against poor risks and unreliable clients. Such restrictions eventually find their natural outlet in the investors, tenants and related business interests who ultimately contribute to the premiums

Finally, and as we pass from anecdote, the JCT major works form of contract deals with insurance in a radical way, in particular by leaving the parties to prescribe precisely their own requirements, avoiding the prescriptive provisions of predecessor forms. The kinds of risk associated with development and construction are prospectively too great, however, for parties to related contracts for them to countenance payment of major claims out of own resources; the same applies equally to investment and the needs of business occupation. Accordingly, without insurance, development and property investment in all their aspects are essentially unfundable and insupportable. No development or construction agreement, no lease or occupation lease omits to deal in some way with insurance, assuming it is properly drawn. From concept through to demolition, development and eventual redevelopment, the entire life of a project is permeated by risk and the need to insure.

## All change in development

It may not be so long before even the following comments will seem severely outdated. The use of all-risks insurances under joint-names policies, to include the employer, the contractor and, as required by interested parties under development agreements, joint venturers and so on, will continue to have its place.

The move towards multi-party insurance, embracing both the principals and the contractor and professionals may, at first sight, seem revolutionary in seeking to mitigate against the kinds of conflict which may prospectively arise between them. More subtly, the concept is evolutionary in seeking to contain risk over a period, including for the longer term whilst periods of limitation ensure continued exposure to construction risk.

It is becoming more common for employers under construction contracts to arrange insurances for development themselves (see above as to the JCT major works form). For example, a ten-year non-cancellable policy has its attractions in providing professional indemnity protection for contractors and other professionals working on a single project. Not only may this overcome the problem for the employer of relying on those he employs to arrange policies on an annual basis (which may become difficult to maintain over time as market conditions change), so also may the beneficiaries of construction warranties be relieved of similar uncertainty. Include a sufficient class of persons e.g. tenants, in the scope of the policy and one has the makings of a far more rational approach to construction risk.

There are obvious parallels with latent defects insurance, which has been around for some considerable time. In its nature, it does not cover the initial insurance risk and those involved in the construction process may or may not benefit from it. For the consumer, developer, investor or tenant, it represents a significant bulwark against the cost of repairing, restoring or strengthening a building if damage is caused by an inherent defect remaining in the structure after completion. Cover available may extend to mechanical and electrical plant, and other non-structural parts of the building.

It appears that latent defects insurance is not as popular as was once anticipated. However demands by investors, tenants, funders or purchasers may create the need for a decision to effect such insurance post-completion, but then it is inevitably more expensive.

The best time to buy latent defects insurance is before construction starts. The insurers' technical auditors and any professional firms engaged for the purpose will have had their input. Whilst it is an added bonus that technical errors can be corrected, and the risk reduced, so also is the risk the better assessed and restrictions in cover, particularly avoidable ones, reduced. In a post-completion scenario, one is dealing with a *fait accompli*, and the potential for a claim may lie unseen until it physically manifests itself. Insurers who have been engaged prior to construction will thus, for example, have the opportunity to review both the proposals for but also, say, the implementation of the provision of foundations, such as the inadequacy of excavations. The placing of concrete may perhaps be subject to error, particularly in the manner and sufficiency of reinforcement. In one instance, the design of a flat slab was perceived to be inadequate so as to exclude a claim for damage due to excess deflection. A year later it had to be taken up and re-laid. Again, basements tend to leak, but insurers' auditors' recommendations may assist avoidance. Also, insufficiency of cladding gives rise to problems, for example that fixings have been found to be wrongly specified or simply inadequate, the cladding has been constructed in a way to permit ingress of water and so on.

From a landlord and tenant point of view, one only has to turn up a raft of cases on repairing obligations in leases, basement water penetration and wall cladding being popular topics, not least on the issue of whether or not remediation falls to be paid under a repairing clause in a lease or whether it was the result of defective construction. And if the latter, of course, and if available limitation periods have expired, a raft of further legal issues arises, such as if the tenant is not liable to make good under his repairing covenant, and there is no contractual obligation on the landlord to make good the defect, can the landlord nonetheless be made liable under his covenant for 'quiet enjoyment' under the lease, and so on.

The point at issue, therefore, is essentially a simple one. Risks attendant upon development, construction and the longer-term investment are inherently bound up. A contractor's all-risks insurance ceases to apply in the case of damage occurring after practical completion: where a design responsibility is included, PI cover will be maintained for a certain period of after completion. The consequences of defects arising post-completion may continue to be an attributable construction risk, but are separate, distinct and separate from the sort of insurance cover associated with completed buildings i.e., the buildings insurance that must be put in place to follow immediately upon practical completion. It follows that investors and occupiers are concerned with the changes in risk which evolve, and with how they may continue to preserve their interests and in particular how these will relate to each other as construction responsibilities give way to their relationships alone. The longer term must be considered from the outset.

## *The nature of risk*

It is helpful to consider risk as a concept, and as applied to a complex property environment, such as a shopping centre. Risk can take many forms, for example, material damage caused by fire: by contrast consider a more natural damage such as corrosion. The more natural, the less insurable. The consequential risk of flood damage following heavy rain is insurable but the loss of trading profits, following a latent defect causing a building to be unusable, would not be covered (which is not to say that consequential loss is not legally remediable, depending on contractual terms). From a lawyer's perspective, that is an over-simplification and so does not entirely reflect the law as hinted at above, and it is important to understand at an early stage that what an insurer will cover can run on separate lines from the legal liability one is trying to protect.

Again, there may be limited cover for gradual pollution, but none for the adverse effects of the deterioration in public opinion that might arise. A particular distinction between what might give rise to a legal liability but not necessarily be insurable is the matter of personal injury. Legal liabilities for injuries to third parties are covered, but those resulting from a deliberate act are not. Again, the political risk of terrorism is now partially insured by a government-sponsored insurer in the UK (Pool Re) but it is almost impossible to secure cover for the effects of government economic policy such as tax changes. And on the subject of financial issues the risks attendant upon an insolvency can be insured, but those resulting from poor marketing decisions cannot. Technical risks are increasingly insurable, such as computer break-down, but obsolescence is certainly not.

Risks may be controllable within an organisation, or they may be attributable, say, to key suppliers, or the risks can migrate, e.g. the impact on a retailer of a defective product even if it can be recalled by the manufacturer.

The costs of remedy can be direct, e.g. repairing the damage, consequential, e.g. the loss of rental income, and indirect, e.g. the loss of goodwill or market share while not being necessarily attributable to a particular event. Add to this the likelihood of a risk materialising and its severity, including the prospects of more than one event coinciding to produce a particular result, and one has a dilemma. The foreseeable, in its nature, can always be taken into account and is not inherently insurable.

### The changing market

The insurance market considers itself to be mature. The traditional risk carriers, the insurance companies themselves are being joined by banks and capital markets, e.g. the rapid development of direct insurers, for own personal needs. In commercial contracts, substantial elements of risk are transferred sometimes to the capital markets via a variety of financial instruments.

Where a landlord gives commitments to his tenants, the certainty of his profitability can be enhanced by putting the variables onto his tenants, such as through service charges. If a code for standard lease terms gains a statutory foothold, then a natural consequence is the imposition of a balance of risk, and the question of insurability follows behind.

### Risk reduction

A commercial contract containing indemnities is clearly intended to reduce the risk of whatever event to be indemnified. The indemnifer may himself need insurance against that event. Many professional firms could never survive claims if they were not insured through their professional indemnity or errors-and-omissions insurance.

Placing reliance on insurance is a risk in itself. Insurance companies can suffer insolvency, just like any other business, and placing insurance cover with an insurer weaker than oneself is not entirely rational. Insurance brokers (should) pay attention to the financial strengths of insurers and keep their clients advised of any relevant deterioration in the market.

Non-lawyers need to remember that because a policy of professional indemnity insurance is a contract of indemnity, it is the insured, who has caused the loss, that is indemnified, not his innocent victim who, in the event of being successful in proving legal liability, may be recompensed simply because the insured has been resourced by his insurer.

Hence the importance of considering the prospective benefits of insurance during limitation periods and the availability, scope and strength of insurance cover. A warranty may yet end up not being worth the paper it was written on, as happened time and again in consequence of the 1990s' recession.

### New risks

New risks continue to emerge, not least technological ones, such as e-commerce. For example, email as a means of communication is hazardous. Companies may find themselves legally committed by the actions of a careless employee. Sensitive information may be unwittingly disclosed, and no disclaimer or claims of confidentiality will help if information simply ends up in the wrong hands. Again, what if reliance is placed on information on a website that is out of date?

The year 2000 had particular significance, and development and construction were hardly unique in that regard. Again, interception of paper records has been joined by internet fraud and hacking and by the bypassing of sophisticated systems. The detailed knowledge of employees both past and present is a risk to a business in itself.

The property world cannot be isolated from any of this and PFI contracts, for example, whilst not just being a source of finance for government-driven projects, also adopt associated risks. Not just the cost of provision but also running costs are transferred to the private sector by PFI. For the government department concerned it means that many of the vagaries of what might otherwise be tenant-borne service charges are fixed.

Legislative changes can have the effect of imposing a measure of risk where none existed before, for example regulations in relation to contaminated land and a landowner's obligation to clean up contaminated sites if the original polluter cannot be traced or has disappeared. Unacceptable risks to human health are also proscribed, but may be susceptible to an insurance remedy.

The bottom line is that disasters happen, and steps must be taken to mitigate the possibility. The prospects of risk may be highlighted in accounts, and directors' responsibilities are enhanced accordingly. A situation which could kill an organisation may arise even before the damage is identified and attempts made to deal with it. Senior management must therefore be committed to the management of risk and to taking responsibility for it. Amongst professional firms who can assist are, of course, insurance brokers. Their own business traditionally focuses on identifying and analysing risks and then facing up to them. The conditionality of a policy of insurance is a natural reflection, and the insured who is subject to such conditions must heed them and comply with them in order to benefit from the policy.

## Insurance of shopping centres

The events of 9/11 have only served to bring into stark relief the need to consider the prospects of disaster on such a scale and what the consequences might be. Insurance issues concerning a shopping centre, whilst complex, are of course smaller in scale, but reflect the protection which is expected by all those who have a share in real estate investment of any kind. One could use other property investments as an example, but shopping centres are so diverse in nature as to be an ideal platform from which to work for present purposes.

The centre itself may be held on a long lease containing insurance obligations. Even if it is held freehold, the relationship with tenants necessarily implies the imposition of insurance obligations and, alongside these, provisions relating to services. Insofar as insurance obligations are wholly upon tenants then perhaps the less an interest the landowner may have. The practicality is that with multi-occupation premises, structurally interdependent, the maintenance and flow of investment returns depends directly on the relationship between the landlord and its tenants. Rents apart, this will be reflected through the maintenance and upkeep passed down to tenants via service charge and the consequences of specific risks which are covered by insurance, such as fire and explosion.

The sorts of insurance provisions one finds in leases today are a relatively modern phenomenon. Insurance companies which include commercial property investments in their portfolios of assets have encouraged this trend. Post-war Britain and, in particular, the emergence of major redevelopment of town centres from the 1960s onwards has evolved into the kinds of shopping centre developments that we take for granted today, including of course purpose-built out-of-town developments. There is thus a fairly well-defined institutional framework for leases which has become the backbone of the institutional investment market; thus shopping centres must be insured and risks managed, including risks which may give rise to a legal liability.

### Insurable interests

It is in the nature of insurance that the financial misfortunes of a few may be supported by the financial resources of the many. The cost of joining the protected club is the premium paid under the policy. Certain characteristics apply to a policy of insurance, not least of which is that the premium should not be considered a wager or bet. For example, the Life Assurance Act 1774, otherwise known as the Gaming Act called the insurance of lives in which the insured had no interest a 'mischievous form of gaming' and sought to stop the practice of insuring an inevitable event, for example, the death of the king, for which the insured would suffer no financial loss. Thus, the insured must have an insurable interest, and not merely stand to make money out the misfortunes of another. The prospect of financial loss is, clearly, an insurable interest. It must also be possible to assess the probability of the damage and the extent of the prospective loss, and on this basis the bargain between insurer and insured may be struck. Of course certain risks may not be insurable at all and for these one must look to central government, such as war and the effects of radioactive contamination. Terrorism is another matter, however, and is considered further below.

Accordingly, the owner of a shopping centre has an insurable interest according to his own interest. Thus, for example, the interest of a freeholder may be so remote, e.g. because the value lies in a lease for 999 years and derives only a nominal rent, he has an interest but his interest is only nominal. Afford him a nearer reversion, perhaps a share in rents too, and his interest is the more tangible. The head tenant's insurance clause under such a lease will require him, say, to keep the shopping centre insured to the full reinstatement value, including landlord's fixtures and fittings, plant and machinery and (if the landlord has such an interest) so many years loss of rent in the joint names of the head lessor and the head lessee against loss or damage by fire and such other insurable risks as the head lessor may from time to time [reasonably] deem desirable and also to maintain insurance for employer's liability and third party risks. The lease covenants at large are intended to preserve the investment environment in which an income can be derived, and insurance is provided in support of this when prescribed perils befall it including loss of rent.

As an aside, loss of rent may not just arise as a result of damage to buildings or to access to premises. Leases themselves rarely so prescribe, but one available line of insurance is cover against tenant insolvency. The availability is circumscribed by the circumstances but, in the case of major investment, enhancement of covenant strength through the use of bonds and insurance is hardly exclusive to the relationship of landlord and tenant.

Insurance clauses in leases usually prescribe that insurance is to be arranged with insurers of repute, or some such wording. The collapse of a number of insurers in recent years highlights the importance. Once cover is lost, so also is the premium, and disgruntled tenants may be asked to pay yet another premium, perhaps more expensive, on top. Commitments which a landlord of occupying tenants has given, no longer protected by insurance, will have to be met out of his own pocket. Retrospective cover may be available where a claim has not yet arisen. Always expect occupation leases to prescribe that the landlord will insure (or that he may if the tenant fails to do so in those relatively rare cases where the tenant is himself allowed to insure).

### Ownership risks

The very fact of ownership and contractual responsibility for management may make the owner, whether head lessee or freeholder, potentially liable to a large number of people and organisations, the tenants themselves, the public at large, contractors and so on. It may be necessary also to indemnify a superior landlord from liability.

Statutory intervention in any event requires certain kinds of insurance to be carried. Employers' liability and in particular the Employers' Liability (Compulsory Insurance) Regulations 1998 contain provisions for compulsory insurance to meet court awards made if workers successfully sue their employers for injuries sustained, through the negligence of a fellow worker or of the employer himself.

This legislative trend is reflected elsewhere, for example, in unconnected areas such as compulsory third-party-liability motor car insurance for personal injury under the Road Traffic Act 1930, government commitments to provide a pension for all under the National Insurance Act 1946 and, of course, the health and safety at work legislation rendering employers liable to employees for breach of statutory duty in addition to any common law liability.

### What needs to be insured?

A pattern thus begins to emerge: the four major aspects of insurance of a shopping centre are first reinstatement of damage to the buildings, landlord's fixtures and fittings and so on; second, the preservation of income both in the form of rent and service charge; third, liabilities incurred under the terms of a head lease if any and, last, but not least, fourth, legal liabilities arising whether under statute or at common law.

So, under a policy of insurance the insured pays a premium to cover a set period, usually twelve months, under which the insurer is to put the insured in the same position as that pertaining before the misfortune insured against (the peril) occurred. The amount of premium paid reflects the nature and number of perils and the likelihood, i.e. the risk of those perils causing financial loss. So, taking peril with peril, the risk of fire is substantially greater than, say, earthquake. It follows that the more the risk is itself reduced, and insofar as it subsists can be accurately estimated, the less should be the amount of the premium.

What the land or the buildings might have cost, or how they are reflected in the books of account of the owning company may bear no relation at all to the insurance value. The issue is essentially one of the cost of putting oneself back into the position one was in before the peril occurred. Deciding upon the amount to be insured against is the insured's responsibility. If he has under-insured that is his problem (see below as to averaging). If he over-insures he will have paid an unnecessary premium. A policy of insurance is simply a commercial contract and thus represents the bargain between the parties to it. Because the contract created by a policy of insurance is a contract of indemnity, it follows that there is no guaranteed payout. If the insured has over-insured, there is still no obligation on the insurer to pay out the total sum insured if the loss or damage suffered is, actually, a lesser sum. An insurer may retain the choice of reinstating himself but actually doing so may be seen as a rare event.

As a first task, therefore, one must identify the items to be insured and the amounts in question. The first of these, of course, is the buildings, say a cost of £50 million to rebuild, to which must be added the cost of demolition and clearance of the damaged structure, say £5 million, and consultant's fees, say £7.5 million, giving an amount insured against destruction or damage by fire and other perils of £62.5 million. VAT may need also to be added, to the extent that reconstruction costs would be liable to non-recoverable tax.

As to the annual gross rent, let us assume, £1 million (valuers need not comment!) that it would require one year for necessary consents plus three years to build in the event of total destruction. Accordingly, four years loss of rent should be covered giving a total of £4 million. This may be seen as generous where one is simply rebuilding in a mirror image something for which planning permission already exists and continues to operate. Consideration must also be given to non-recoverable items such as service charge expenditure which continues to be incurred notwithstanding, such as additional marketing costs, legal fees and so on.

Many occupation leases also make provision for early termination if reinstatement has not been effected within a certain period. Loss of rent should be assessed to ensure that there is sufficient cover to allow completely reletting the centre once rebuilding has been completed and that the adverse effect on turnover rents is also taken into account.

Public liability insurance is essential to protect owners, managers and staff who may be sued by members of the public. The amount of indemnity needs careful assessment based not only on risk of death or injury to people in and around the centre but also for damage to third-party property and the consequences that might result. The figure may be substantially increased if the centre is, say, adjacent or near to other large buildings where there is a real risk of fire spreading from the centre as a result of negligent action. Public liability cover is usually arranged for a specific limit of indemnity on each and every loss, but subject to these limitations the aggregate may be unlimited.

As to employer's liability, mention has been made of the Employer's Liability (Compulsory Insurance) Regulations 1998. Insurers now generally impose a limit of £10 million on the indemnity provided for each incident. It may therefore be considered appropriate to arrange additional layers of cover to reflect the exposure created by the numbers of staff who could be involved in a major incident. There is a statutory obligation to display the appropriate certificate of insurance.

The management may also effect insurance affording all centre staff personal accident cover should they be injured or killed in the course of their duties.

### Shopping centre perils

Even in the best-run centre, potential disasters abound and can be divided into three broad categories: first, loss or damage resulting to the property including the consequential loss of income; second, liabilities arising from owning, occupying and managing the centre; third, perils affecting the person.

As to property, the most common peril is fire. Historically, it was the earliest peril to be generally insured against. Other perils used to be termed 'full special perils' and indeed some leases used such terminology, borrowing from insurers. They rarely do now. These include things like lightning, explosion and spontaneous combustion; the social perils of riot, civil commotion and malicious damage; the natural perils of earthquake, flood, storm, tempest, etc. and miscellaneous risks such as aircraft and objects falling therefrom, and damage caused by burst waterpipes and tanks, impact of vehicles, etc. To this must be added, today, in particular, vandalism and it is important that insurance covers both riot and malicious damage.

Today, whatever the lease terms (but beware inability to recover the cost of premiums from tenants insofar as the lease does not specify) fire and special perils are mostly replaced by a wider-ranging commercial all-risks policy. Property investment relies of course on an income stream, and so loss of rent becomes more important. Further, loss of service charge, through cesser clauses in leases, requires cover so that payments can continue to be made to cleaning contractors, maintenance specialists and so on whose contracts will likely contain a minimum period for notice of cancellation.

Consider also common parts and other non-lettable parts such as management offices and contents, which may comprise a separate item in the policy schedule.

### Engineering perils

Apart from defects liability considerations, and the need for PI cover during a limitation period prescribed by warranties, an important aspect of ongoing buildings maintenance is compliance with statutory requirements. Some forms of lease can be criticised for failure to include engineering cover and, as part of the services, timely inspections.

Quite apart from the practical necessity of protecting assets, there are also legal duties to ensure that all statutory inspections are completed on time and that remedial work is undertaken within the required timescales. Prosecution by the environmental health officer or the Health and Safety Executive, where personal injury is involved, may follow. Due inspection of plant and machinery may mitigate against civil liability. Regulations include the Lifting Operations and Lifting Equipment Regulations 1998, the Pressure Systems Safety Regulations 2000, the Control of Substances Hazardous to Health Regulations 2002, the Electricity at Work Regulations 1989 and the Provision and Use of Work Equipment Regulations 1998. Expect insurers to avoid claims where requirements have not been met. Subject to this, engineering perils (which can relate to both mechanical and electrical engineering plant) includes sudden and unforeseen breakdown of insured plant and any impact damage to surrounding property owned by the insured. Engineering insurers will normally carry out the statutory inspections of plant, at six-monthly intervals for lifting equipment and annually for boilers and pressure plant. Insurers may cover other equipment such as escalators which fall outside statutory requirements but for which an independent inspection would be prudent.

### Other perils

Sprinkler leakage comes to mind, but one might just as well add theft, fidelity and contingency risks, accidental breakage of glass and loss of licence. The parameters of cover offered must be examined carefully. Tenant rent default is another area which has been mentioned above.

Fidelity and contingency insurance focuses on the reliability and honesty of the numerous people involved in running a centre, providing services to tenants, handling money and so on. There will be many opportunities for theft of cash or property, but so long as all these people are personnel strictly under the manager's control or at least for whom he has responsibility, a blanket fidelity policy may be effected. This is a typical example of an insurer not being prepared to be 'selected' against, by insuring only those individuals who have the most opportunity of committing crime. Insurers discourage selection.

Legal indemnity insurance, such as in respect of a restrictive covenant, or defective title insurance will likely as not have been effected when the land was bought for the particular development. That is not to say that the need might be perceived later, say upon disposal to a new owner or on refinancing, but in its nature it may also be reflected in the price paid for the investment, and it would be an unusual circumstance if any at all in which the cost could be passed down to tenants, for the clear inference is that not one solicitor picked up the problem along the way.

By contrast, particularly in retail shopping centres where tenant mix is important to overall success, leases of licensed premises should, in the interest of the income stream, require tenants to insure against loss of licence.

Finally, replacement of glass in shopfronts is a responsibility usually placed upon tenants under their leases, and so they should be required to insure accordingly.

The second important area is liabilities, i.e. those that the owner may incur by reason and in consequence of owning, occupying or managing the centre and employing staff for the purpose. All staff accidents at work should be insured against as should civil liabilities that may be incurred to members of the public for injury or damage arising from defects and shortcomings, and also from lack of care, e.g. in cleaning or maintaining the centre. Quite possibly an accident may give rise to a blend of issues, some of them construction based.

Mention should also be made of personal liabilities. Where the provision of services includes the provision of staff and the cost of providing those staff, that cost may include employee benefits: pensions, the consequences of potentially hazardous tasks, ill health and private medical treatment, permanent health insurance. Some of these costs may be reflected in payroll deductions, but others may be seen as additional.

### Preservation of income

In a shopping centre, or indeed any property investment, supporting the income stream is paramount. Suggestions above as to the period of loss of rent insurance, for example, must now be given greater meaning by prescribing the amount of insurance cover required. Whilst existing planning may suffice if the centre can be rebuilt identically, the need for planning advice should not be ignored. If demolition would be a breach of planning and, say, planning policies have moved on and a replacement building would not be desirable or practical in the original form, then the timescale for redevelopment may be substantial indeed. Break clauses may more readily be exercised and, without suitable insurances in place, the income stream may be severely impaired or simply dry up. Development and investment are inextricably intertwined.

Another factor in loss of rent is the prospect of rent review, perhaps even more than one review, during the cesser period, and so the amount to be insured against may, actually, need to be greater than the sum of the passing rents and other commitments. Turnover rents add further complexity, as also VAT insofar as it is not recoverable.

Mention has already been made of inclusion of certain service costs where these continue to be incurred after damage or destruction. Some costs may even increase, e.g. security, and without adequate insurance the landlord may end up incurring the cost personally. An interesting situation may arise where the damage or destruction can be attributed to professional negligence in construction, with the possibility of buildings insurers exercising their rights of subrogation to pursue those who have given warranties, they in turn falling on their PI insurers (if they can), another blessing for the lawyers.

Where loss of rent insurance is to include the prospects of increase of rent on review, centre owners may be reluctant to ball-park the prospects via their insurance arrangements, for fear of forewarning tenants so that they can develop their arguments. The landlord could, of course, insure the prospects of increased rent separately and at his own expense, but a more practical alternative is for the main policy to provide automatic cover for reviews, however they may turn out, subject to a maximum percentage uplift, without the need for additional premium.

### Reinstatement

After a period of reducing interest rates, one must entertain the possibility, at least, of inflation, and of the prospects of rebuilding cost increasing during the twelve-month period from commencement of a policy. Allowance should be made for increases in cost over the relevant period of planning and reconstruction, say four years as a yardstick.

Statutory requirements increase and, where major reinstatement is required and also requiring planning permission, there is the prospect of difficulty. Not that those with a knowledge of planning law will have failed to recall that planning permission for development in one location cannot impose conditions affecting development in another, hence the resort to statutory agreements from time to time, but that qualification may prove difficult to apply in a confined context where the two are so inextricably bound up that the planning site may need to be more widely defined.

Another distinction to be made is whether any modernisation or upgrade is simply desirable or necessary. Tenants seeking to minimise their service charge exposure should look closely to the terms of their leases and to the propriety of additional charges being made upon them for services following reinstatement. These prospects require specific legal advice. Where major reconstruction presents difficulties for the landlord in recovery of expenditure in the future, it may yet be a matter of compromise and a variation to the lease. After all, the trading location may be ideal and the tenant's right to terminate may not be his best answer. Of course, if it is worth his while holding his landlord to ransom, that is another matter. A landlord's express right to terminate if reinstatement cannot be achieved in the same form or within a certain period may provide the key, and so the fault line may lie in the original negotiation of the lease rather than the practicalities that the landlord would like to achieve.

When estimating costs of reinstatement, inflation is better catered for on a compound basis, and although inflation rates cannot necessarily be predicted, a small margin of error in addition may be beneficial.

Neither should VAT be forgotten, particularly the regime pertaining to commercial buildings since 1989. Provision for total rebuilding must therefore allow for VAT where the status is otherwise exempt, partially exempt or even non-registered. A modern shopping centre, by contrast, will usually see transparent recoveries. Where non-retail units are concerned, however, the VAT status of occupiers in some circumstances is inimical to the developer's option to tax and it is wise, in the case of mixed-use development not only for the developer to be to be advised on tax for the purpose of his own recoveries but, in turn, to ensure that his buildings reinstatement insurance is sufficient.

Almost every aspect of maintenance of an investment somehow interfaces with or is inextricably bound up with the initial construction, and the change from contractor's insurance to buildings insurance along the way is an event, not a watershed.

Finally, before leaving the subject of how much insurance, a word is required about 'average'. As has been seen, over-insurance will do the insured no good, a policy of insurance is a contract of indemnity, not a wager or a bet and the maximum obligation of the insurer is the true amount of the loss insured against. But if the insured under-insures, what then? Under an average clause the insured is consequently responsible for a percentage of any loss in proportion to the degree of under-insurance. Hence the provision in occupation leases, on the one hand that the landlord will insure to the full reinstatement value but, on the other, that any shortfall will be made up out of his own monies. Inevitably, that shortfall will usually be qualified by an exception to the extent of something along the lines of the act, default, neglect of the tenant, its servants, licensees or invitees.

### Who should insure?

Developers undertaking new development will have been closely involved in the detailed negotiation of head leasehold structures, funding and so on. Arranging insurance for the completed development, and ongoing during its life will no doubt be delegated in succession to people who, over time, will have had no connection at all with the initial process. Whilst it is unlikely that a centre manager will be charged with the task of ensuring that all necessary insurances are in place, whoever has the task delegated to him must look to the owner's responsibilities to head landlords, to funders and to occupation tenants respectively. Inescapably, this means reviewing principal documents lest there be breach of contractual commitments, taking legal advice again if necessary rather than struggling to get the right answers.

Whilst much can be expected, exceptions abound. Perhaps even the head landlord may have retained responsibility for insurance, passing the cost of premiums down to the head tenant who may distribute them again amongst individual tenants. Again, the head lease may specify, if not a particular insurance company then a superior landlord's choice of insurer or broker through which the insurance is placed. The motivation may include the need to secure insurance from a reputable company, insurer's commission or the business needs, say, of the head lessees or funder's own company if they are insurers in their own right.

Whatever the insurance requirements, that is not to say that an alternative quote may not be obtained: an interested party may need to be convinced and then agree expressly an alternative course. At all events, costs are there to be controlled, and a proactive approach maintained.

### Tenant's insurance

A tenant will in any case need to insure its own fixtures and fittings, stock and so on. Plant and machinery within the demise and belonging to the tenant may be the subject of a strict requirement on the part of the landlord to insure for fear of damage to other parts of the development.

The fabric of the premises and fittings or improvements which form part of the landlord's fixtures are all insurable by the landlord as such (unless the tenant is to insure buildings himself). Where, as in most cases, the landlord insures, what happens, for example, if there has been cesser of rent occasioned by one or more insured risks, the period of reinstatement has taken longer than the cesser period, and either the tenant has no right to terminate or, having such a right, he elects not to do so? In that case, a tenant should consider insuring against resumption of rental payments and, possibly, this cover will form part of the overall business interruption insurance that the tenant may arrange to protect his business.

A few other points should be noted, the first of these indeed being a tenant's often expressed desire that his interest be noted on the landlord's insurance policies. Noting an interest is not the same as making the tenant a joint insured. Under what is known as a general interest clause the insurer may thus acknowledge the interests of all parties having a legal interest in the property. A tenant's demand to be made a joint insured needs to be considered with caution, however. At times, this may be justified, for example where the full market rent is not reserved, or the tenant has perhaps paid a premium so that (in the absence of a proviso for re-entry upon an insolvency event, which is the legal catalyst) the tenant has a capital interest in the lease. As a joint insured the tenant may thus become a joint loss payee: he will have the right to approve expenditure on making good damage. The landlord's interest may be impaired, say, because his co-insured, the tenant, has failed to disclose material facts, in which case cover under the whole policy may be impaired.

If the landlord alone insures, a sensible arrangement, for the greater comfort of the tenant, may be to provide in the lease for the landlord to seek, and if he can then obtain a waiver of the insurer's recovery (subrogation) rights. The principle is that if the loss or damage is attributable to the tenant then, in pursuance of such rights, the insurer may seek to recover against the person responsible for the loss or damage. In the case of a tenant reimbursing the landlord for the insurance premium, case law has proved beneficial to tenants and so insurers are usually willing to provide a blanket endorsement affording a waiver in favour of all the tenants in a property in respect of their units and also the common parts.

For fear that the above may be too much of an over-simplification, advice must be sought in specific instances as they arise.

Non-disclosure of material facts by tenants as joint named insured is not the only problem of its kind. All named insured must make appropriate disclosures including not least the landlord. Hopefully the policy will contain a landlord's clause which ensures that the landlord's position is not prejudiced by any act or default of the tenant of which the landlord is not actually aware. When he becomes aware he must disclose and pay any relevant increase in premium (and if the occupation lease so prescribes, he may make recovery in turn from his tenant).

Another area of concern, particularly in shopping centres, is the consequence of tenants mounting displays or exhibitions in the malls which, of course, form part of the common parts. In this sort of circumstance, a separate insurance, perhaps covering only the event in question, is desirable, better still if it is effected with the same insurer which insures the centre at large.

### Contractor's insurance

Apart from tenants' own insurances, there are others to take into account as well, particularly those of contractors engaged, if not in reinstatement, then in alterations, extensions and so on. Contractors' insurance should be reviewed with care, not only in relation to the works to be carried out, but particularly in relation to public liability. It is not just a matter of members of the public or traders suffering injury or loss, but regard should be had to the prospects of being sued by tenants under the covenant for quiet enjoyment in their leases. This presents a dilemma which goes beyond insurance because whilst most of the consequences of the work may be insurable, if the work is to be carried out in a certain way, indeed with the approval of the centre management or whoever the employer under the contract is, and if the works themselves give rise to a landlord's liability for breach of the covenant for quiet enjoyment, having been carried out entirely in a manner which is satisfactory to the landlord, then perhaps insurance will not be the answer. Indeed, in such circumstances, the tenant's business interruption insurers may be chasing after the landlord under their own subrogation rights. Most modern occupation leases protect the landlord with suitable exceptions and reservations, but not all do, so never make assumptions: seek legal advice.

Voluntary lapses apart, for which insurance may not be the answer in context, ensuring that contractors carry the appropriate kinds as well as levels of insurance is thus an inherent part of good risk management.

Having railed against the dangers of joint names insurance, one cannot ignore the fact that this is precisely how the JCT standard forms (other than major works) approach insurances relating to works and existing structures, so that it is the employer who is to take out and maintain a joint names policy in respect of the existing structures and for the full reinstatement value of the works. It is wise to review existing buildings and other insurances to see what the impact of additional building works may be on the efficacy of such insurance. The principles of disclosure apply to all insurance, and if there is a distinction to be made at all the joint names involved alongside the contractor in new development are largely removed in a physical sense.

Finally, works carried out within the confines of an existing property may offend other legal rights, e.g. rights of access, rights of way and so on. In some cases such interruption may yet amount to breach of the covenant for quiet enjoyment and regard must be had to related occupation leases and the specific terms of those leases including exceptions and reservations (see above). It is one thing to reserve to a landlord a bundle of rights to ensure that maintenance and remedial work can be carried out without causing legal offence to tenants. It is another thing entirely if the proposed works are not contemplated by leases and if express consents may be required.

## Premiums

Under a contract of insurance, payment of the premium is the insured's part of the bargain. Strictly, the premium should be paid before cover commences, but policies can and commonly do provide for commencement of cover so long as the premium is paid within a certain period, following which cover may be withdrawn. In insurance parlance this is a policy warranty, under other kinds of contract a condition precedent or, as the case may be, a condition of which time is of the essence. Insurers demand that, in any event, terrorism premiums be paid within thirty days of inception cover, and they in turn will pay the Pool Re.

Where buildings insurance is to be apportioned between tenants, the amount of the premium may be made part of the service charge and divided according to the service charge provisions of the leases, or it may be separately identified and charged separately. If it is charged by way of rent, the significance of 'rent' lies not in the separation of the charge but in the remedies available to the landlord in case of tenant default. Indeed, service charges at large can also be reserved by way of rent. In particular, the saving provisions section 146 of the Law of Property Act 1925, and the court's discretion in equity to afford relief from forfeiture may be swept aside if the landlord wishes to use non-payment of 'rent' as a grounds for obtaining possession. The landlord's insurance covenant may correspondingly lie in the covenant to provide services or, separately, to insure, to which is added an obligation to reinstate (save for vitiation by the tenant, see above).

As regards uninsured losses, these may be intentional, the result of failure to insure or failure of the insurer itself, e.g. upon insolvency. The institutional market is sensitive to such issues, having long adopted the format of the 'clear' lease, affording the relative certainty of an identifiable constant rent. This, of course, lies at the heart of the investment. The respective obligations of landlord and tenant, blended with the safeguards of insurance, are intended to bring about a certain result. Obligations which the lease omits to place upon the tenant may revisit the landlord in the form of a breach of the covenant for quiet enjoyment, even if the landlord has not himself covenanted to perform the duty in question.

## Claims

Curiously, development agreements and leases usually have little to say about the making of insurance claims. They may go as far as imposing compliance with the terms of a policy, but one does not expect, for example, to see obligations upon an occupation tenant to co-operate with his landlord by providing information and assistance to facilitate the making of a claim.

Perhaps legal documents should flesh out such matters in more detail, and the point is debatable. Notions such as the common law duty of care meanwhile come to mind, not forgetting also expected provisions as to non-vitiation of insurance.

Insurance claims require a measure of practicality and common sense, however. Whoever has the responsibility delegated to him, say the property manager, must carefully log all relevant instances, gathering as much evidence as possible, statements, reports, photographs and so on. The insurer will have its own reporting procedures and provide the necessary forms. Care is required to ensure compliance with policy conditions.

In an environment such as a shopping centre, claims by the public require particular care and in the case of personal injury, insurers should be notified without delay. A good relationship with an insurer's claims department is desirable and, although it goes against the lawyer's grain, from an insurance perspective, it is desirable to encourage the insurer to appoint a loss adjuster to handle claims or to delegate claims settlement to an appointed manager. Large shopping centres can expect a trickle of incidents, and good protocols can only make for good management. Of course, and again one may not expect to find it mentioned in legal documents, regular consultations with interested parties are desirable, tenant meetings and so on, to review matters. A measure of understanding and personal confidence may prove particularly helpful in the event of a major claim.

## Terrorism

The events of 9/11 are not the only ones to provide interest in terrorism. The bombing activities of the IRA, particularly in London, naturally provoked insurers to impose exclusions relating to fire and explosion caused by terrorist activities during the early 1990s. Some limited cover remained but times have moved on. To tackle terrorism a government-assisted insurer was created, called the Pool Reinsurance Company Limited, known in the business as 'Pool Re'. A particular feature of Pool Re cover is that the insured cannot select risks. So, if you have a number of properties covered by the same insurer, you simply cannot decide that one only of these may be selected. Pool Re insures only UK risks although terrorism insurance has been much influenced by the events of 9/11.

From 1 January 2002 insurers excluded terrorism damage entirely. The result was that by July 2003 Pool Re announced that it would meet the wider exclusion. From August 2002 terrorist cover was authorised to extend from fire and explosion to an all risks basis including biological and chemical contamination. From 1 January 2003 nuclear contamination has been added. War risks are still excluded, even from Pool Re cover. Additional exclusions are added for hacking and virus damage to electrical components, but the basis of Pool Re is that the protection it affords is still confined to 'acts of persons acting on behalf of or in connection with any organisation which carries out activities directed towards the overthrowing or influencing by force or violence of Her Majesty's Government in the United Kingdom or any other government *de jure* or *de facto*'. On the other hand, insurers exclude 'the act of any person or groups of persons whether acting alone or on behalf of or in connection with any organisation or government committed for political, religious, ideological or similar purposes including the intention to influence any government or put the public or any section of the public in fear'. This overlap thus lays bare a lacuna in the availability of terrorist protection.

Since 1 January 2003 the operation of Pool Re has changed so that each insurer will be responsible for a proportion of an aggregate insurance market layer of cover, increasing from a maximum of £60 million in any one year in 2003 to £200 million in 2006. The primary insurers will set their own premiums for this element of cover. The limited capacity for very large risks means inevitably that the cost is high and increasing.

The wider cover available is important, risks such as biological or chemical damage being particularly significant to investments such as shopping centres, with ready access to members of the public. Nonetheless, it appears that there is little appetite on the part of government to widen the definition of act of terrorism for Pool Re purposes, and it appears that the insurance market has yet to soften its own stance universally, although some insurers are prepared to give wider cover.

The whole subject of terrorism insurance is a developing one which deserves constant attention by investors and the taking of suitable professional advice.

## Conclusion

As has been seen, insurance has its place as an important element in the management of risk: in its nature it is far from synonymous with management of risk. Its underlying necessity, however, has necessarily resulted in the Chancellor taxing it and, as at 2003, the insurance premium tax (IPT) on gross premiums stood at 5 per cent. Inspection fees for plant and machinery do not attract 'IPT', but of course being fees they attract VAT instead.

Both the scope and limitations of insurance highlight the elements of risk attendant upon all development agreements, construction agreements and the leases which go to make up the resultant investment. The assistance kindly provided in sourcing material for this chapter also provokes thoughts about the drafting of the legal documents associated with the development process.

Lawyers are naturally conservative: however good one's commercial legal drafting, knowledge and comprehension of the subject matter are essential pre-requisites of adequate documentation. What is necessary to deliver a desired result, and to ring-fence undesirable exposure, is risk management at work, although risk management professionals may not always see it that way. For the surveyors and other professionals involved in the technical, financial and commercial activities and negotiations leading to lawyers' instructions, it behoves both the instructors and the instructed to broaden their horizons and to visit every proposition as if it were novel. The reality of legal experience is that legal safeguards are the product of practical safeguards but that, after many years of experience it remains that the practical safeguards are, unhappily, largely a product of the imagination based upon that experience, and not always the inspiration of the client. Never assume that you have the answers, the probability is that we are light years away from them and it is only good fortune that the risk has not visited us before we are adequately prepared. Better check your PI cover!

# Index